Fundamentals and Applications of Solar Energy—Part II

Ihab H. Farag and Stephen S. Melsheimer, *editors*

W. R. Aiman
B. N. Baron
R. W. Breault
E. C. Clausen
M. A. Darwish
R. R. Davison
B. G. Eldridge
M. M. Elsayed
D. W. Gregg
H. E. Hager
W. B. Harris
S. Lakshmanan
J. C. Ludlow
F. K. Manasse
V. K. Mathur
E. Kenneth May
B. E. Poling
D. G. Raynor
R. E. Rocheleau
T. W. Russell
D. A. Sama
A. H. Shah
K. J. Sladek
B. E. Sparks
S. Sunderan
I. S. Taha
P. F. Thompson
C. B. Thorsness
S. Tostengard
V. Venkataramanan

AIChE Symposium Series
Number 210 1981 Volume 77

Published by

American Institute of Chemical Engineers

345 East 47 Street New York, New York 10017

American Institute of Chemical Engineers
345 East 47 Street, New York, N.Y. 10017

Library of Congress Card Catalog No 80-16305

ISSN 0065-8812 ISBN 0-8169-0218-6

Printed in the United States of America by
Lew A. Cummings Co., Inc.

FOREWORD

The search for alternative energy resources in the present time of dwindling supplies represents a thought provoking and growing field for the engineer. Harnessing the unlimited resource of the sun is a challenging alternative that has attracted the interest and talents of many engineers and scientists, and which the AIChE has recognized through increased programming at recent technical meetings.

This is the second AIChE Symposium Series Volume in the area of solar energy. It is based upon papers presented at the 89th National AIChE Meeting, August 17-20, 1980, in Portland, Oregon and the 73rd AIChE Annual Meeting, held November 16-20, 1980, in Chicago, Illinois, in two solar energy sessions chaired by the editors. With the continuation of AIChE programming in Solar Energy, it is the hope of the editors that Volumes dedicated to reporting the contributions of Chemical Engineers in this area will continue.

The papers collected in this Volume cover a wide range of fundamental and applied research including thermal storage, solar cells and generation, industrial applications, manufacturing and biomass utilization. Thus, it reflects the broad range of research in solar energy applications, rather than being focused in any specific area.

The editors wish to thank the individuals who contributed to bringing this volume into existence. The cooperation of the authors and reviewers in meeting the tight schedule that led to the timely printing of this volume is especially appreciated.

Ihab H. Farag, *editor*
Assistant Professor of Chemical Engineering
University of New Hampshire 03824

Stephen S. Melsheimer, *editor*
Professor of Chemical Engineering
Clemson University
Clemson, South Carolina 29631

REVIEWERS

W. R. Aiman
J. A. Bailey
S. Dasgupta
D. J. Graves
D. W. Gregg
G. D. Gupta
T. G. Lenz
S. Lakshmanan
M. S. Peters
H. I. Robison
C. F. Tsang

CONTENTS

PHASE CHANGE FLUIDS FOR SOLAR THERMAL SYSTEMS

This paper explores the use, for the storage of solar energy, of phase change materials which are suspended or emulsified in an immiscible carrier fluid. Emulsions of up to 50 weight % paraffin wax in water were found to be very fluid, highly stable, and quite flame resistent. Such easily pumped emulsions allow for an increase in stored energy density while avoiding the severe heat transfer rate problems normally encountered with phase change storage. Since the suspended phase change materials can be used both to collect and store solar energy, a heat transfer step is eliminated and the energy may be stored at a higher average temperature. This in turn results in a higher thermodynamic availability which is shown to be particularly advantageous in the storage of solar energy for refrigeration or heat pump systems.

D. A. SAMA
and
K. J. SLADEK
Department of Chemical Engineering
University of Lowell
Lowell, MA 01854

Active solar heating systems generally use water or air as working fluids. Water and crushed stone have been used for thermal storage, and commercial storage units containing salt hydrates have recently become available.[1] This paper explores novel uses of phase change materials in solar heating and refrigeration.

Phase change systems can be advantageous in either the collection or storage of solar heat. For example, the heat transfer coefficient in a collector is increased considerably by using ammonia or a halogenated hydrocarbon which vaporizes in the collector. Such fluids can also avoid problems of freezing in the collector. Ammonia, for instance, has a freezing point of -77.7^{o}C and would not freeze in a solar collector even in extreme northern locations. However, phase change systems which result in a gaseous phase are not well suited to energy storage since the volume of the gaseous storage system is prohibitively large. Thus boiling or subliming systems are not attractive from an energy storage viewpoint.

Solid-solid phase change systems are possible but, in an active solar system, it is difficult to transfer heat into or out of them.[2] Thus the logical choice for a phase change system from the energy storage viewpoint is a solid-liquid change.

Numerous solid-liquid phase change systems are available.[3] Their main disadvantage is low heat transfer rate... particularly during heat removal where a solid phase encrusts the exchanger surface.

Our proposal is to incorporate the solid-liquid phase change material as small particles in an immiscible carrier fluid. This of course dilutes the phase change effect but enhances dramatically the heat transfer rates into and out of the phase change material which may be emulsified, in suspension or a slurry. Thus we have in effect a dual-phase phase change system. An additional advantage of such systems compared with traditional storage systems is that they can be used directly in the collector as well. This eliminates the heat exchanger normally required to transfer heat from the collecting fluid into the storage medium.

0065-8812-81-4473-0210-$2.00

This in turn allows the heat to be stored at a higher temperature.

Two obvious categories of phase change fluids are mixtures of a suitable wax in an aqueous phase or mixtures of an appropriate salt hydrate in an immiscible nonaqueous fluid. By using suitable emulsifiers and mixers it should not be difficult to make stable emulsions of either category. These would typically consist of micron-sized particles stabilized by high molecular weight surfactants. After extended use the particles might tend to grow and eventually the emulsion might have to be replaced or regenerated. Regeneration could be accomplished by including a mixer or homogenizer in the collector fluid loop. In this case less emulsifier would be needed and larger particles could be tolerated. A system of this type might be a coarse dispersion aided by a detergent. A mixer located at the collector exit would resuspend the components for the next pass.

In the present paper we report on a phase change fluid consisting of a stabilized emulsion of paraffin wax in water. Paraffin waxes are particularly good choices because they are low in price, are available in a wide range of melting temperatures, are noncorrosive, and are readily emulsified. A possible disadvantage of a paraffin is its flammability. However, this is mitigated by the presence of the continuous water phase which surrounds each wax particle.

EXPERIMENTAL

Experimental work on making and characterizing stable paraffin wax in water emulsions has been rather more suggestive than definitive and will be presented in very little detail here. Recipes for wax in water emulsions can be obtained from many suppliers of emulsifiers. Equipment for making dispersions includes mixers and blenders, colloid mills, and homogenizers of various types.[4] We have found that emulsions containing up to 50 wt % wax are not difficult to prepare. These are very fluid and are stable for many, many heating-cooling cycles through the wax melting point. Our current research plans include a study of the long range effect of cyclical heating and cooling on the stability of such emulsions. The diameter of the emulsified wax droplets is ½ to 4 micrometers depending on the energy intensity expended in dispersing the wax initially. Thus, blenders result in droplets at the large end of the scale, while homogenizers give the smallest droplets. Emulsions having the larger droplets show some creaming after a month's storage, but this is easily reversed by gentle mixing.

Since water is the continuous phase, these emulsions tend to be nonflammable. That is, we were not able to ignite them with a match flame. To ignite a wax in water emulsion would probably require an ignition source of sufficient strength to vaporize all the water out of the emulsion.

We have not yet measured the heat transfer characteristics of these wax in water emulsions. It is clear, however, that the micron sized wax particles will offer little resistance to heat transfer in either the melting or cooling cycle. The two-phase emulsion behaves in effect as a single phase fluid of very high heat capacity. Note that in scaling up the quantity of the emulsion there is no increase in the resistance to heat flow into or out of the individual wax particles. This is in contrast to a one component phase change material which, like the proverbial dinosaur, finds heat transfer more difficult as it increases in size.

Figures 1 and 2 give some thermal properties of one emulsion containing 40 wt % wax. These were obtained using a DuPont Model 1090 Differential Scanning Colorimeter. The wax (Fig. 1) shows a rather broad melting transition starting at 47°C. The thermal behavior of the emulsified wax (Fig. 2) is virtually identical to that of the wax itself.

In summary, one class of suspended phase change materials---paraffin wax in water emulsions---is not difficult to prepare in concentrations up to 50 wt % wax. These emulsions are highly stable, very fluid, and appear to have thermal properties that are the sums of properties of their constituents.

In the following section, we develop the theory of suspended phase change materials. Rather than using measured properties of a particular emulsion, we will use a hypothetical 50 wt % emulsion of sharp melting wax in water. The thermal properties of the emulsion are assumed to be the sum of the thermal properties of the wax phase and those of the water phase.

THEORY

We will evaluate, from both weight and volume points of view, the storage requirements of three idealized solar heating systems using water, wax and a 50 weight percent wax in water emulsion. In all cases a total of 1.055 gigajoules (10^6 BTU) are collected between temperatures of 32.2°C (90°F) and 71.1°C (160°F). This heat storage is equivalent to approximately 1/6 Bbl of oil or about 7 gallons.

The systems are compared in Table 1. The thermal properties chosen for the wax are typical of paraffins (3). The melting point of the wax was assumed to be 65.6°C (150°F) and its heat of fusion 209.3 kilojoules/kilogram (90 BTU/lb). The pure wax of course results in the largest storage savings, about 50%. The wax-water emulsion saves about 1/3 of the storage weight and 1/4 of the storage volume compared to water. There is, however, a more significant thermodynamic difference between the systems.

Though each of the storage systems in Table 1 contains the same amount of heat, and are therefore equivalent by the First Law of Thermodynamics, they are quite different according to the Second Law. The heat addition to the wax systems is at higher average temperatures than to the water system. This results in a higher thermodynamic "availability", B, where:

$$B = H - T_oS \tag{1}$$

Availability is a property which is not conserved, and the net effect of higher availability depends on factors extraneous to the storage system. The manifestations of higher availability in heat and steam are discussed in detail by Sama (5). Higher availability may manifest itself as:

1 - A reduction in fuel requirements to produce a given amount of work.

2 - The ability to transfer heat to a higher temperature level.

3 - A reduction in heat exchanger area.

Availability may also be "expended" without any benefits at all.

Let us look into the availability differences between the systems in Table 1. The increase in availability of a heat storage system can be calculated from the heat capacity and the latent heat of any phase changes. The change in availability due to sensible heat addition is:

$$dB = dH - T_o dS \tag{2}$$

where T_o is the minimum absolute temperature at which the storage system can provide heat. Substituting for the change in enthalpy and entropy as a function of heat capacity, mass, and temperature difference we get:

$$\int_{T_o}^{T} dB = \int_{T_o}^{T} mC_p dT - T_o \int_{T_o}^{T} \frac{mC_p dT}{T} \tag{3}$$

Assuming that the heat capacity, C_p is constant between T_o and the melting point:

$$\Delta B = B_T - B_{T_o} = mC_p \left[(T - T_o) - T_o \ln \frac{T}{T_o} \right] \tag{4}$$

Taking the availability at T_o as zero, the availability, at any temperature up to the melting point is:

$$B_T = mC_p \left[(T - T_o) - T_o \ln \frac{T}{T_o} \right]$$

$$\text{for } T_o < T < T_M \tag{5}$$

At the melting point, the change in availability is

$$\Delta B = m \left[\Delta H_{fus.} - \frac{T_o \Delta H_{fus.}}{T_M} \right] = m \, \Delta H_{fus.} \times \left[\frac{T_M - T_o}{T_M} \right] \tag{6}$$

Assuming that the heat capacity of the liquid phase C_p', is constant, the availability change for sensible heat addition above the melting point is:

$$\Delta B' = mC_p' \left[(T - T_M) - T_o \ln \frac{T}{T_M} \right]$$

for $T > T_M$ (7)

Equations (5), (6), and (7) were used to calculate the availability change of the wax and water storage systems of Table 1. The availabilities are plotted in Figure 3 as functions of heat addition. The slope of these curves dB/dQ at any point is the Carnot efficiency for the increment of heat added at temperature T or $(T-T_o)/T$. During sensible heat addition T is increasing, hence the slope increases. During the phase change heat addition, when the wax is melting, the temperature is constant hence the slope is constant. The portion of the curve for wax where melting occurs is therefore a straight line.

Note that though the total heat addition is the same, the availability of the wax is nearly 40% greater than that of water.

The availability of the 50% wax/water system lies exactly half way between the wax and the water curves. Thus the 50% wax-water system, in addition to saving 1/3 of the storage mass, has about 20% greater availability than the water system. Unfortunately this increased availability is not effectively utilized in the ordinary solar heating system. It manifests itself as a larger ΔT during the heat release process which, depending on the heat transfer coefficient compared to that of pure water, may or may not result in smaller heat exchangers.

Better use would be made of the wax or wax/water systems if their energy storage were being utilized to do work such as to power a solar refrigeration system or a heat pump. Suppose for example that we need 1.055 GJ of stored heat above 60° C (140° F) to operate an absorption refrigeration system or a heat pump. For this case the heat is stored between 60°C and 71.1° C (160°F). The relative storage requirements for the water, the wax and the wax/water systems are listed in Table 2. Note that water requires about 5 times more storage capacity than 100% wax and about 3 times more capacity than the 50% wax/water system. The relative water storage requirements for other cases will increase as the temperature range for storage decreases and vice-versa.

Though Table 2 indicates that 100% wax has a superior stored energy density than the wax in water emulsion, we must remember that this is an idealized comparison. In an actual system, the wax chosen for the emulsion can have a higher melting temperature than the wax chosen for the 100% wax system. This is due to the fact that the emulsion can be used directly in the collector. In such a case, for a given quantity of stored heat, the availability stored in the emulsion will be greater than that stored in the 100% wax. Thus in applications where the stored solar energy is used for its thermodynamic work potential, the stored energy density of the wax emulsion may even be superior to that of 100% wax. This is in addition to the enhanced heat transfer rate capabilities of the emulsion over those of 100% wax.

The conclusion is clear. The use of dual-phase phase change fluids offers significant advantages in the storage of solar heat ....particularly for solar refrigeration or solar heat pump systems.

NOMENCLATURE

- B Thermodynamic Availability, J
- H Enthalpy, J
- T Absolute Temperature, °K
- T_o Minimum Temperature at which heat can be transferred to ambient, °K
- T_M Melting Temperature of Wax, °K
- S Entropy, J/°K
- C_p Heat Capacity, J/kg°K
- Q Heat added to storage system, J

LITERATURE CITED

1. Parkinson, Gerald, "Phase-Change Materials Aid in Heat Recovery," Chem. Eng., July 16, 1979, pp. 52-58.

2. Galloway, T. R., "Paraffin Wax Heat Storage for Solar-Heated Homes," U. S. Dept. of Energy Report on Contract No. W-7405-Eng-48.

3. Hale, D. V., M. J. Hoover, and M. J. O'Neill; "Phase Change Materials Handbook," NASA CR-61363, Sept., 1971.

4. Rees, L. H., "Evaluating Homogenizers for Chemical Processing," Chem. Eng., May 13, 1974, pp. 86-92.

5. Sama, D. A., "Looking at the True Value of Steam," Oil & Gas Journal, April 14, 1980, pp. 103-119.

Table 1

Comparison of Idealized Heat Storage System per 1.055 GJ (10^6 Btu) of Storage for Solar Heating.

	Water	Wax	50 wt% Wax/Water
Outlet Temp., ^oC (^oF)	71.1 (160)	71.1 (160)	71.1 (160)
Inlet Temp., ^oC (^oF)	32.2 (90)	32.2 (90)	32.2 (90)
Melting Point, ^oC (^oF)	N.A.	65.6 (150)	65.6 (150)
Heat of Fusion kJ/kg (Btu/lb)	N.A.	209.3 (90)	104.7 (45)
Heat Capacity Below M.P. kJ/kg^oC (Btu/lb^oF)	4.187 (1)	3.768 (0.9)	3.977 (0.95)
Heat Capacity Above M.P. kJ/kg^oC (Btu/lb^oF)	4.187 (1)	2.931 (0.7)	3.559 (0.85)
Specific Energy Storage J/kg (Btu/lb)	162.8 (70)	351.2 (151)	243.1 (104.5)
Density at Outlet Temp. kg/m^3 (lb/ft^3)	977.1 (61)	781.7 (48.8)	879.4 (54.9)
Storage Mass, kg (lb)	6,480 (14,286)	3004 (6623)	4340 (9569)
Storage Volume, m^3 (ft^3)	6.632 (234.2)	3.843 (135.7)	4.936 (174.3)
Fraction of Water Mass	1	0.464	0.670
Fraction of Water Volume	1	0.579	0.744

Table 2

Comparison of Storage Requirements for Solar Refrigeration/Heat Pump System Requiring 1.055GJ (10^6 Btu) of Heat above 60 C.

	Water	Wax	50 wt% Wax/Water
Outlet Temp., ^oC (^oF)	71.1 (160)	71.1 (160)	71.1 (160)
Inlet Temp., ^oC (^oF)	60 (140)	60 (140)	60 (140)
Specific Energy Storage J/kg (Btu/lb)	46.52 (20)	246.6 (106)	146.5 (63)
Storage Mass, kg (lb)	22,680 (50,000)	4,279 (9,434)	7,200 (15,873)
Storage Volume, m^3 (ft^3)	23.21 (819.7)	5.474 (193.3)	8.186 (289.1)
Fraction of Water Mass	1	0.189	0.317
Fraction of Water Volume	1	0.236	0.353

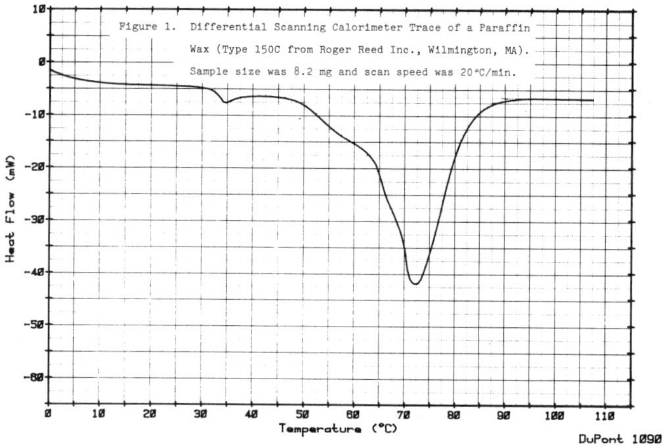

Figure 1. Differential Scanning Calorimeter Trace of a Paraffin Wax (Type 150C from Roger Reed Inc., Wilmington, MA). Sample size was 8.2 mg and scan speed was 20°C/min.

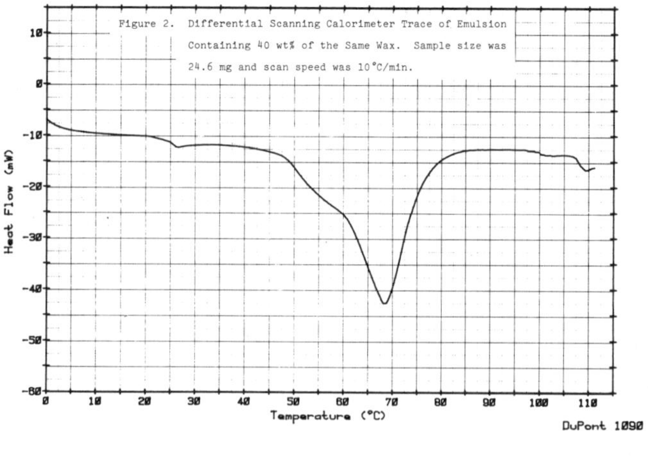

Figure 2. Differential Scanning Calorimeter Trace of Emulsion Containing 40 wt% of the Same Wax. Sample size was 24.6 mg and scan speed was 10°C/min.

Figure 3. Availability Versus Heat Addition

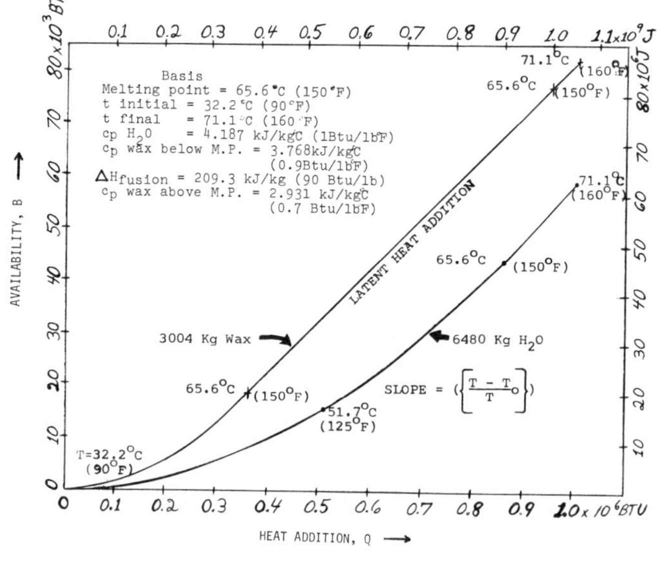

STORING SOLAR ENERGY BY LIQUID PHASE DIELS-ALDER REACTIONS

Using chemical reactions to store energy is not a new concept. But the idea of using chemical reactions entirely in the liquid phase at low temperatures for relatively small scale storage applications, i.e., for space heating, is relatively undeveloped. In this paper, the method of how chemical reactions could be used to store energy without generation and storage of new phases is described. Criteria a reaction must satisfy to successfully store energy are presented and several Diels-Alder reactions are evaluated in terms of these criteria. At this time, no reactions have been identified that meet all the criteria.

B. G. SPARKS
P. F. THOMPSON
and
B. E. POLING
University of Missouri-Rolla
Rolla, MO 65401

Energy storage is an essential part of many solar energy systems used in home heating. Common storage methods, shown in Figure 1

Rock Beds

Water Tanks

Phase Change

Phase Change With Separation

Figure 1 Common Energy Storage Methods Ranked in Order of Increasing Complexity and Storage Capacity

may be evaluated in terms of both their energy storage capacity and their complexity. The simplest method is at the top of the list (rock beds), while the most complex (phase change with separation) is at the bottom. The energy storage capacities of these different storage methods are characterized by the values shown in Table 1.

Unfortunately, schemes that have the highest storage capacities are also the most complex. For a new scheme to gain acceptance as an energy storage method, it would, for example, have to be as simple as heating and cooling water, while possessing an energy storage capacity greater than water.

Table 1 Energy Evolved in Some Common Processes

Process	Energy Evolved $JX10^{-8}$
Cool 1 m^3 of water from 365° to 305°K	2.51
Solidify 1 m^3 of Glauber's salt	3.66
Freeze 1 m^3 of water	3.34
Condense steam to give 1 m^3 of water	22.60

Chemical reaction schemes proposed to date have not satisfied these criteria. Those reactions that have been examined generally have involved multiple or gas phases, frequently accompanied by separation and storage of reactants and/or products, and high temperatures. Such schemes have extensive equipment requirements-pumps, heat exchangers, compressors, and storage vessels-and require high temperature solar collection. These schemes do have potential for long term energy storage, and the thermodynamics of both the forward and reverse reaction can be controlled by adding or removing certain key reactants. But the complexity of these systems renders them unsuitable for small

0065-8812-81-6038-0210-$2.00

scale applications.

If the forward and reverse processes in a chemical reaction could alternately be favored, and if all species were liquids, always remained in the same vessel, and were always at relatively low temperatures, the objections commonly encountered to chemical reaction storage could be avoided (Poling, 1979 (1)). How this might be accomplished can be explained by referring to Figure 2.

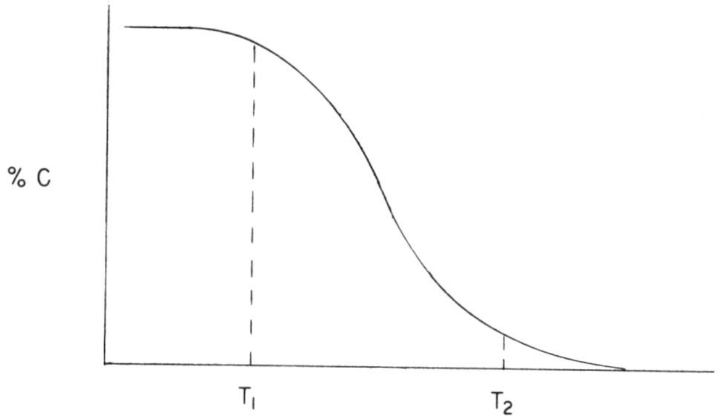

Figure 2 Equilibrium Composition vs. Temperature for the Exothermic Reaction, A+B → C. Both the Heat of Reaction and Sensible Heat Must be Supplied to Heat from T_1 to T_2

which shows the equilibrium composition versus temperature for the exothermic, reversible reaction, A+B → C. At T_1 which might be about 305 K (90 F), an equilibrium A-B-C mixture would be predominately C. At T_2, which might be about 365 K (200 F), the mixture is primarily A and B. Thus, the energy required to heat the A-B-C mixture from T_1 to T_2 would be not only the sensible heat, but also the heat required to convert the C to A and B. Furthermore, if A, B, and C were liquids between T_1 and T_2, storing energy by heating an A-B-C mixture would have the same relatively simple equipment requirements as storing energy by heating water. Questions that remain are how much more energy could the reaction mixture store than just heating water? And do compounds exist that demonstrate the behavior shown in Figure 2?

The energy storage capacity of a hypothetical reaction can be calculated with the equations in Figure 3

$$Q = M\,C_p\,\Delta T + (-\Delta H^\circ)(N)$$

$$\frac{d \ln K_x}{dT} = \frac{\Delta H^\circ}{RT^2} \qquad K_x = \frac{X_C}{X_A X_B}$$

N = moles of C at T_2 − moles of C at T_1

Figure 3 Equations to Calculate Energy Storage Capacity for A+B ⇄ C

and physical properties in Table 2.

Table 2 Typical Physical Properties for A and B

	ρ, kg/m^3	MW	C_p, kJ/kg K
A	940	68	.44
B	11300	98	.44

An equimolar A-B mixture contains 13600 gr moles per m^3

The heat storage capacity, Q, that results for particular sets of thermodynamic properties is shown in Table 3.

	$-\Delta H^0$, kJ mol^{-1}	K_x AT 305^{0}K FOR A+B⇄C	N/6800	Q, Jx10^{-8}
1.	0	-	0	1.25
2.	40	8*	.46	2.50
3.	80	37*	.75	5.32
4.	120	192*	.89	8.51
5.	120	.01	.0005	1.29
6.	120	10^6	.005	1.64

*K_x THAT MAXIMIZES N.

Table 3 Heat Stored When 1 m^3 of a Reaction Mixture is Heated from 305°K to 365°K for Different Values of ΔH° and K_x

A reaction with a heat of reaction of 80 KJ mol^{-1} (19 Kcal mol^{-1}) could have an energy storage capacity twice that of water. Table 3 also illustrates that there are three characteristics a reaction must possess to successfully store energy. It must have a high heat of reaction, an equilibrium constant such that a large composition change occurs in going from T_1 to T_2, and the reaction mixture must be concentrated in reacting species. The heat of reaction plays a dual role. Not only does it appear in the calculations directly, but it also determines how much the equilibrium composition changes with temperature. As for the equilibrium constant, there will be some best value that optimizes a reaction's energy storage capacity. If the equilibrium constant is too big or too small, either the reactants or products may be favored over the entire temperature range and no composition change will occur (see lines 5 and 6, Table 3). Finally, if the reacting species are dilute, the number of moles reacting, and thus the energy stored will be small.

Diels-Alder reactions can be used to store energy (Poling, (1), Lenz and Hegedus, (2)) and we have examined several of these reactions. Diels-Alder reactions (see Figure 4)

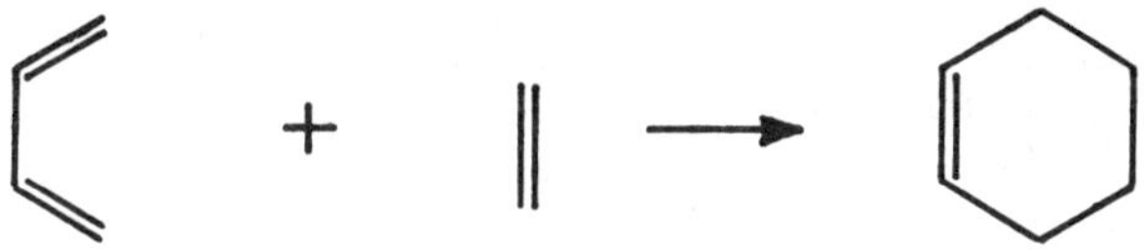

DIENE DIENOPHILE ADDUCT

Figure 4 Typical Diels-Alder Reaction

are addition reactions between two compounds termed a diene and dienophile. Typical heats of reaction are 70 to 80 kJ mol^{-1} (17 to 19 kcal mol^{-1}). Since a great variety of compounds can serve as dienes and dienophiles, it is not unreasonable to expect that some combination would possess the desired equilibrium constant. Diels-Alder reactions have the advantage of being fairly clean reactions and they can be catalyzed, so that equilibrium compositions should be fairly easily attained. Current efforts are directed towards identifying which, if any, Diels-Alder reactions can successfully store energy. Unfortunately, thermodynamic data for Diels-Alder reactions at the desired conditions are scarce. It has become apparent that to identify desirable reactions it will be necessary to characterize the effect of concentration, solvent, and substituent on heats of reaction and equilibrium constants. The role played by substituents and concentration will become more apparent with discussion of some example reactions.

One reaction that is reversible, and for which some data are available is the dimerization of cyclopentadiene (Laidler, (3)) shown in Figure 5.

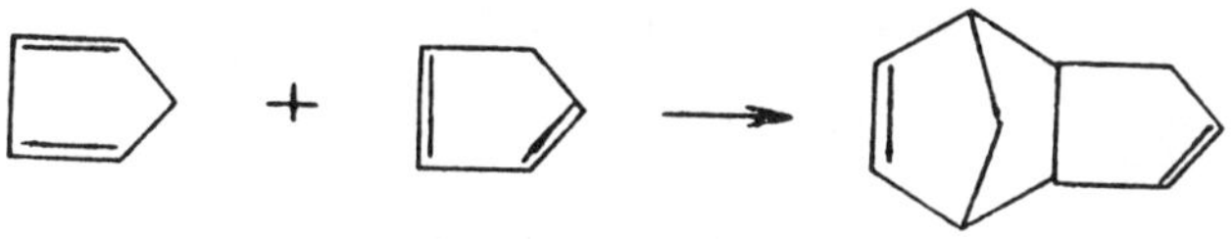

$\Delta H° = -77.4$ KJ/mol

$\Delta S° = -185$ J/°Kmol

$$T^* = \frac{\Delta H°}{\Delta S°} = 554°K = 537°F$$

Figure 5 Thermodynamic Properties for the Dimerization of Cyclopentadiene (Laidler, 1965)

This reaction involves liquids and has a high enough heat of reaction, but the equilibrium constant is too large at the desired temperature (T*, the temperature where K = 1ℓ/mol, is about 550 K). The reaction between furan and maleic anhydride, is also reversible (Anet, (4)). We have measured thermodynamic properties for this reaction and the results are shown in Figure 6.

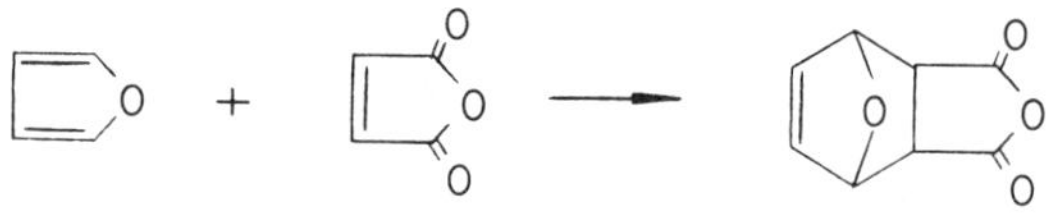

FURAN(I) MALEIC ANHYDRIDE (II) ADDUCT

T, °C	CONCENTRATIONS, mol/l x 10^4		K_x x 10^{-5}
	I and II	ADDUCT (Exo Form)	
40.4	1.09	8.91	18.4
50.2	1.61	8.39	8.0

$\Delta H° = 17.2$ Kcal $T^* = 656°K = 383°C$

Figure 6 Results for the Furan-Maleic Anhydride Reaction

The heat of reaction remains sufficiently high, but the equilibrium constant is still too large. If one replaces the furan with 2-methylfuran, the equilibrium constant goes down (Seltzer, (5)) as shown in Figure 7.

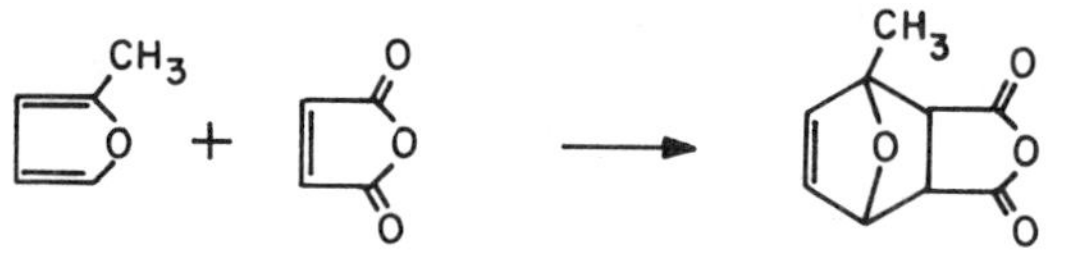

2-METHYLFURAN (I) MALEIC ANHYDRIDE (II) ADDUCT (III)

AT 50°C, $\frac{C_I C_{II}}{C_{III}} = 1.7$ moles/ℓ

Figure 7 Equilibrium Constant for the Reaction Between 2-Methylfuran and Maleic Anhydride (Seltzer, 1963)

Our measurements support Seltzer's findings in that significant changes in the equilibrium composition are observed with changes in temperature. Thus is appears that the ability to modify substituents will allow the desired equilibrium constant to be obtained.

The decrease in K when furan is replaced by methylfuran could result either from a decrease in ΔH°, which would be bad for energy storage, or an increase in ΔS°, which would be good for energy storage. This raises the question of whether it is even possible to lower K without decreasing ΔH°. Data of Vaughan and Andersen (6) indicate that it is. They studied the reaction between anthracene and various maleic anhydrides and some of their results are shown in Table 4

Table 4 Heats of Reaction and Equilibrium Constants at 142°C for Reactions Between Anthracene and Substituted Maleic Anhydrides (Vaughan and Andersen (1956)(6))

	ΔH°, kJ/mol	K, ℓ/mol
$C_4H_2O_3$	—	70,000
C_4HBrO_3	—	34,000
$C_4HCH_3O_3$	157	229
$C_4HC_3H_7O_3$	205	64.5

When methyl maleic anhydride was replaced by n-propyl maleic anhydride, the equilibrium constant went down, and ΔH° went up. Increasing the size of substituent had the same effect on K as in the furan-methylfuran case, and a favorable effect on ΔH°.

Another variable which can have a large effect on the equilibrium constant is the concentration of the reacting species. T* for the dimerization of cyclopentadiene is 749 K when the reaction occurs in a paraffin solvent, but 554 K when the reactant is pure liquid cyclopentadiene. This indicates that the equilibrium constant for adduct formation in dilute solution is much greater than for the case where there is no solvent. In the design of energy storage systems where concentrated solutions are desired, caution must be exercised in using equilibrium constants measured in dilute solutions.

This paper has developed the concept of using liquid phase chemical reactions to store energy, and several Diels-Alder reactions have been discussed, primarily in terms of their energy storage capacity. To successfully store energy, a reaction must have a high heat of reaction, the right equilibrium constant, and the reacting species must be concentrated. To date, no reaction has been identified that satisfies all three of these criteria. As the behavior of reversible Diels-Alder reactions becomes better characterized, it is hoped that such reactions will be identified. But demonstration of a high heat storage capacity will not, by itself, result in a reaction being used to store energy. Other factors that will have to be considered are costs, stability, toxicity, and corrosivity of the chemicals. In a chemical storage system, savings will be realized due to a smaller storage tank and improved collector performance. We estimate this will allow spending 15 to 20¢ per lb. for the chemicals. The question of chemical stability is perhaps the most crucial one. Replacement of the chemicals every year or two, for example, would be unsatisfactory. Thus it will be essential to identify those conditions that avoid undesirable side reactions.

But in spite of these potential problems, the idea of storing energy by liquid phase chemical reactions seems to be a good one. The method avoids many of the problems commonly associated with chemical reaction storage, and offers an operationally simple method for increasing energy storage capacities over existing methods.

ACKNOWLEDGMENT

The authors gratefully acknowledge the Department of Energy, Appropriate Technology Program, for financial support.

NOTATION

C_p Solution heat capacity, assumed constant with T and composition

$\Delta G°$ Change in Gibbs Energy on reaction, kJ/mol

$\Delta H°$ Change in entropy on reaction, kJ/mol K

K Equilibrium constant in terms of concentration

K_x Equilibrium constant in terms of mole fraction

M Mass of solution

Q Heat storage capacity

R Gas constant

T Temperature, K

T* T where K=1

x Mole fraction

ρ Density

LITERATURE CITED

1. Poling, B. E., "Using Reversible Liquid Phase Reactions to Store Thermal Energy", Proceedings of Solar Energy Storage Options Workshop, San Antonio, March, 1979.

2. Lenz, T. G., and L. S. Hegedus, "Moderate Temperature Solar Thermochemical Systems-An Application of Diels-Alder Chemistry", International Solar Energy Storage Options Workshop, San Antonio, March, 1979.

3. Laidler, K. J., Chemical Kinetics, McGraw-Hill, New York (1965), see p. 200.

4. Anet, F. A. L., Tetrahedron Lett, 1219 (1962).

5. Seltzer, Stanley, J. Am. Chem. Soc., 85, 1360 (1963).

6. Vaughan, W. R., and Andersen, K. S., J. Org. Chem., 21, 673 (1956).

DESIGN AND ECONOMICS OF A SOLAR HEATER AQUIFER STORAGE SYSTEM

Water bearing formations or aquifers provide promising, long-term storage for solar energy at the cost of two or more water wells. This paper combines aquifer simulation with solar heater and short-term surface storage simulation in a parametric economic study of aquifer-solar energy systems. Aquifer storage-solar heater systems are shown to have considerable cost potential in large, space heating applications. As with conventional systems, the heater area is still the principal cost. The use of short-term surface storage is shown to be cost effective with larger ratios of heater area to well flow rate.

S. TOSTENGARD
W. B. HARRIS
and
R. R. DAVISON

Department of Chemical Engineering
Texas A&M University
College Station, TX 77843

In spite of large increases in energy costs, the economic case for solar energy has not been overwhelming, not even for low temperature applications. A major reason is that solar energy is diffuse and intermittent, requiring large areas and an auxiliary heating system. In building space heating, which is potentially the most attractive solar application, heaters sit idle most of the time. Their use is limited primarily to clear, cold days during which only a small fraction of the annual available solar energy can be collected. More energy is available in summers, and with warm weather it can be collected more efficiently, but normally, except for some use for hot water heating, it is wasted.

Nothing can be done about low solar intensities, but the other problems could be greatly lessened by inexpensive storage that permitted hot water produced on an annual cycle to be used as needed. The benefits would be great even without a probable three to five fold increase in energy collection. Lower-cost, single glazed heaters could be used, as most of the energy would be collected in warm weather. Heat collected on an annual basis is predictable, largely eliminating stand-by system use, and diminishing the fear that wide spread use of solar energy would cause serious load imbalance for the utilities.

The use of aquifers for large-scale, low-cost solar energy storage appears to be an order of magnitude less expensive than storage alternatives.

The goal of this work is to show the interaction of solar heater, short-term surface storage, and aquifer storage with weather, aquifer and cost parameters in determining capital costs. Two locations were used corresponding to solar and climatic conditions of central New Mexico and North Dakota. Most aquifer variables were assumed constant except flow rate, depth and formation thickness. The solar-surface system and aquifer were independently simulated and then combined to obtain system costs.

AQUIFER STORAGE

Aquifer storage has been described previously (1to 5) but the significant facts will be reviewed. Good aquifers are very widespread, and in many large populated areas good water wells are found anywhere. The depth and capacity varies as does the water quality, and in many areas several aquifers are available.

An aquifer is a porous, permeable geological formation, usually sand and gravel, filled with water under sufficient head to cause a good flow into a well. Most aquifers have very little movement, and stored hot water will still be in place months later.

By the nature of aquifers, the stored volumes must be large, at least in the

0065-8812-81-4309-0210-$2.00

millions of gallons. The system depends on size and the thickness of the aquifer and overlying formations to achieve high efficiency. The formation is initially cold and recovery is not good the first year. It rises with each cycle, however, and after three to five years it can be over 80% for an aquifer having a thickness of at least fifty feet.

For the system to operate efficiently with minimal environment and water chemistry problems, a well doublet system (4) should be used. In warm weather, water from the supply well is withdrawn, heated and injected into the storage well. In cold weather, the flow is reversed. Hot water is withdrawn from the storage well, sent to buildings for heating, and the return water is reinjected into the supply well.

A regenerative effect occurs in each well so that most of the heat stored in the formation is recovered, and there is no net withdrawal of the return water, and after a few cycles, heat is no longer wasted heating the cold aquifer water to a useful level.

SURFACE SYSTEM DESCRIPTION AND SIMULATION

The configuration of the surface system is shown in Figure 1. This arrangement is arbitrary to a considerable extent. The surface storage serves to extend aquifer use by permitting peak heater production to considerably exceed the well pumping rate.

Circulation is from cell No.1 to the heater as soon as the heater shows a net gain. The flow is regulated to achieve a 13°F temperature rise across the heater. Return flow from the heater enters whichever cell is just cooler than the return temperature. When flow from the heater reaches the preset minimum injection temperature, injection begins at the circulation rate or the maximum well flowrate, whichever is smaller. When the circulation rate exceeds the well flow rate, the excess goes to storage. When injection begins, water is withdrawn from the supply well at the same rate and sent to Cell No.1.

For large heater areas, the storage temperature continues to rise, and the average injection temperature considerably exceeds the minimum injection temperature. Heater circulation ceases as soon as no net gain is possible but injection continues until cell No.6 falls below the minimum injection temperature. For large heater areas and surface volumes, this can be all night.

TABLE 1
Solar Heater Assumptions

No. glazings	1
Glazing emissivity	0.88
Plate absorption and emissivity	0.95
Transmittance, absorption product	0.84
Wind speed	3 mph
Bottom insulation	3 in.
Insulation thermal conductivity	0.03 Btu / ft-hr-°F
Edge loss	negligable

TABLE 2
Aquifer And Formation Properties

Aquifer:	
Porosity	0.3
Heat capacity*	0.5215 Btu/lb.-°F
Density*	109.72 lb/ft
Thermal conductivity	1.775 Btu/hr-ft-°F
Formation:	
Heat capacity	0.308
Density	126 lb/ft^3
Thermal conductivity	1.3 Btu/hr-ft-°F

*Property = property water X porosity + property sand (1 - porosity)

The simulation used for this system proceeds in a series of steady state steps. Flow from cell No.1 to the solar heater is assumed constant at the final temperature in cell No.1 from the previous step. With an assumed constant flow rate and solar input, the heater output is also constant during the time step. The heater flow is then injected into the proper cell which is assumed to be completely mixed. Each cell contains one sixth of the total storage. With a step by step procedure through the cells, the temperature in each cell at the end of the time period is calculated.

The simulated solar heater was a single glazed flat-plate collector with properties shown in Table 1. The upward heat losses were calculated from the equation developed by Klein (6). The overall optical efficiency was assumed to vary with sun angle as indicated by Hottel and Woertz (7). Solar input was calculated from the ASHRAE Subroutine SUN which gives clear sky radiation at any location and time. Clear sky results were modified by a cloud factor calculated from the ratio of measured monthly values of radiation (8) to clear sky values.

To get the amount of heat collected for a month, the surface system was simulated for four consecutive identical days corresponding to the 15th day of the month (see Table 3). To calculate solar heater losses, average monthly daylight temperatures were used for each location.

A variety of runs were made over a range of heater angles with the optimum always found near 30° in New Mexico and 40° in North Dakota. These angles, used throughout this work, are very roughly latitude minus 5°

AQUIFER MODELING

The aquifer was modeled by dividing the formation radially into a series of rings and vertically into slices. Each segment thus formed was assumed to be at a uniform temperature at any time. The aquifer was divided into 3 layers. The formation above and below the aquifer was assumed impermeable. The overlying formation was divided into two 10ft thick slices next to the aquifer, then one 20ft slice and two 50ft slices. Radially the formation was divided into 25 rings. Beyond the outer ring and last slice, the temperature was assumed constant at the initial uniform aquifer temperature. Bottom losses were assumed to equal upward losses.

The aquifer properties are given in Table 2. During pumping, uniform flow was assumed in all three aquifer slices, with the water in each segment constituting a uniformly mixed volume. Also during pumping, radial heat transfer was assumed to occur by fluid flow only. Conductive heat transfer was calculated by assuming uniform heat transfer between segment centers using the distance between centers, the boundary area, the temperature differences between segments calculated from the previous time increment and the thermal conductivity. Assuming steady state time periods, heat balances were made progressively from the well bore, always assumed at injection temperature during injection, to the periphery and from the center of the aquifer outward.

This is a simplistic model, ignoring as it does fluid convection, and assuming pseudo steady state with uniform temperature and complete mixing in each segment. This may underestimate edge losses which can be very dependent on convection, but this was minimized by limiting the study to large injection volumes. On the otherhand, losses calculated by this model can be high because of a mathematical diffusion that results from a series of completely mixed segments. Any change at the well bore is immediately reflected, however slightly, at the edge. The results actually agree fairly well with more sophisticated models, and for a parametric study such as this, the errors are not very significant.

COMBINED SYSTEM

In order to comine the aquifer simulation and solar system simulation, some operating variables affecting each must be fixed. In all cases it was assumed that return water from the heated buildings wat at 85°F and that it was reinjected into the supply well at this temperature. Figure 2 shows the average temperature versus reciprocal time of the water that is subsequently withdrawn from the supply well and sent to the solar heaters. The solar heater simulations were all done assuming 80°F feed water, and results in each case were corrected for the small differences shown in Figure 2. The temperature after 4 years was used for the correction. It was also assumed that when the temperature of the water recovered from the storage wells fell below 95°F, pumping ceased.

Calculations of aquifer storage efficiency was based on the energy recovered above 85°F with the withdrawal temperature equal or above 95°F, thus

$$\text{Efficiency} = \frac{\Sigma\, Q_R\,(T_R - 85);\ T_R \geq 95}{\Sigma\, Q_I\,(T_I - 85)}$$

where Q_R and T_R are values of the total quantity of water recoverd and the average recovery temperature for each recovery time period. Q_I and T_I are values of the quantity of water and average temperature for each injection time period.

In the northern region all aquifers were assumed to be originally at 50°F, and in the more southernly region, they were assumed to be at 60°F. Figures 3 to 5 show values of aquifer recovery efficiency as functions of aquifer thickness, initial temperature and average injection temperature. These must be combined with a corresponding heater simulation for total system results.

For simplicity it was assumed that there are five cold months and seven warm months. During the seven warm months, all of the produced hot water is injected into the aquifer. During the five cold months, the water from the solar heater is assumed to be used immediately along with water from the aquifer. In actuality, even in winter there

TABLE 3

EXAMPLE OF SOLAR SIMULATION OUTPUT
NEW MEXICO CLIMATE, 50,000 FT^2 HEATER, 120° MINIMUM INJECTION TEMPERATURE
30° HEATER TILT, 80^oF WATER FROM AQUIFER, 200 GPM WELLS

MONTH	RESULTS OF CONSECTIVE DAY RUNS IN FT^3/DAY				FT^3/MONTH (LAST 2 DAYS)	AVERAGE TEMPERATURE	BTU'S COLLECTED PER MONTH $X10^{-9}$
	1	2	3	4			
APRIL	15579	16081	16079	16079	489,131	136	1.557
MAY	16940	16939	16939	16939	515,300	137	1.672
JUNE	17710	17709	17709	17709	538,715	138	1.782
JULY	17696	17697	17697	17697	538,331	137	1.747
AUGUST	16995	16994	16994	16994	516,941	139	1.742
SEPTEMBER	16056	16057	16057	16057	488,467	138	1.615
OCTOBER	14182	14176	14176	14176	431,230	133	1.292
NOVEMBER*	11235	11221	11219	11219	341,283	126	0.873
DECEMBER	8151	8694	8694	8694	264,463	122	0.611
JANUARY	9474	9516	9520	9521	289.614	123	0.687
FEBRUARY	12594	12609	12610	12610	383,598	128	1.029
MARCH	14263	14274	14274	14274	434,278	133	1.301
*WINTER MONTHS					TOTAL HEAT COLLECTED		15.906
					SUMMER MONTHS CORRECTED FOR AQUIFER EFFICIENCY		14.160

will be times when excess water is produced by the solar heaters, and at these times the water would be injected. Storage times would be so short, however, that aquifer loss would be small having little effect on annual energy production.

Figure 6 shows the net summer (April-October) collection in total Btu's at the New Mexico location for various heater sizes, well flow rates, and surface storage/area ratios. It also shows the average injection temperature for a minimum injection temperature of 120^oF. The collected energy has been corrected for aquifer efficency by using the proper graph, Figure 3 in this case. The computer simulation produces a monthly volume and average temperature, and these are combined to get total injected volumes and an overall average injection temperature. An example of these data is shown in Table 3. To get the total annual heat collected, the amount collected during the five cold months must be added to the results in Figure 6.

The cost assumptions are given in Table 4. The cost parameters have been combined in three cases as shown. Two solar heater costs were used, although the effect of size was not included. Only one surface storage cost was used, but this is the least significant item. The cost parameters were combined to obtain the fixed capital investment for the heater-storage system per million Btu's stored annually. The distribution system cost, which can vary greatly, is not included, thus

$$I = \frac{A(C_A + C_S \text{ X Ratio}) + C_W \text{ X D X 2}}{\text{A X Q}}$$

in which

- A = heater area in $ft.^2$
- C_A = heater cost in $\$/ft.^2$
- C_S = storage cost in \$/gal.
- C_W = well cost in \$/ft.
- Ratio = ratio of storage volume to heater area
- D = well depth in ft.
- Q = annual net energy recovered in millions of Btu's per square foot of heater

I = Capital investment per million Btu of annual capacity

Operating costs will be a function of I plus distribution costs. The pumping cost is relatively small, though not insignificant, but it can vary greatly with the water level and distribution system.

TABLE 4
COST PARAMETERS

Case	Heater cost, $/ft.^{2}	Storage cost $/Gal.	Well Depth,Ft.
A	10	0.75	500
B	7	0.75	1000
C	10	0.75	1000
Well Flow Rate in Gal/Min	200	500	1000
Well Cost $/Ft.	75	90	100

To obtain the operating costs derived from these numbers as shown in Table 5, the capital cost was assumed to be borrowed for 25 years at 12%, the annual maintenance was taken as 5% of capital, and a pumping cost of $0.50 per million Btu's was assumed. Annual operating cost would then be

$$\text{Capital Cost} = \left[(\text{Cost}/10^6\text{Btu}) \times (10^6\text{Btu/yr})\right] \times \frac{0.12(1 + 0.12)^{25}}{(1 + 0.12)^{25} - 1}$$

$$\text{Maintenance} = \left[(\text{Cost}/10^6\text{Btu})\ (10^6\text{Btu/yr})\right] \times 0.05$$

$$\text{Pumping} = (10^6\ \text{Btu/yr}) \times 0.50$$

where the value of (Cost/10^6Btu) is taken from the appropriate figure 7 to 13. If it is possible to take advantage of the 40% investment tax credit for individuals, the capital cost would only be 60% of the above.

RESULTS

Some results of aquifer simulation are shown in Figures 3 to 5 at various aquifer conditions and injection temperatures. As would be expected, the efficiency goes up with formation thickness, provided sufficient volume is injected, with temperature and with total volume. As pointed out earlier, the program used to obtain these results ignores stratification caused by density differances between the stored water and the original aquifer water. This can cause significant loss at the boundaries, and is best countered by injecting large volumes.

In Figure 6 it can be seen that for a given well capacity, large heater areas result in higher average collection temperatures because of the inability of the well to accept the volume of hot water produced by the solar heaters. It also can be seen that this result can be considerably lessened by short-term surface storage. The higher temperatures result in lower collector efficiency which is not offset by a somewhat higher aquifer efficiency. As a result, the net heat delivered per unit area tends to decline.

Figures 7 to 13 show system costs for a variety of well parameters, heater area to storage volume ratios, and cost prameters.

For ratio's of area (ft.^{2}) to well flow (GPM) less than about 250 there appears to be little gain or even loss from surface storage. This results primarily from the fact that the storage does not result in the collection of more heat, as shown in Figure 6, exacerbated somewhat by the added cost of storage. For larger area to well capacity ratios, the optimum surface storage to area ratio increases with increasing area. Any generalization is affected by relative costs of wells and area. As shown in Figures 7 and 8, the optimum surface storage increases with well cost and somewhat with heater cost.

Figure 9 shows the effect of minimum injection temperature on cost. As stated earlier, system efficiency declines with increasing temperature because of a decline in heater efficiency. It should be noted, however, that the cost of distribution has not been included. The cost goes down, of course, with increasing well capacity, Figure 10, depending on the relative cost of heater and wells. The cost will also go down with increasing aquifer thickness as shown in Figure 11, assuming that the total injection volume is sufficiently large.

Figure 13, for the northern climate, shows much increased cost due to less heat collected. Also at this climate, none of the areas are large enough to benefit significantly from surface storage. In spite of costs nearly double those in New Mexico, the relative gain from the aquifer storage system would be larger because of the very poor winter performance at these latitudes.

To gain some insight into what operating costs might derive from these numbers, consider the example in Table 3 with Case C costs. From Figure 7 the capital investment is found to be $47.70 per million Btu of annual load. The total system load is 14 X 10^9 Btu. Assume 140 homes, each with an annual load of 10^8Btu. The capital investment for each home would be $4,770. If the distribution system added $1,000 to this, the cost would be $5,770 per house, but in many cases, this could be offset by the saving of a furnace.

Considering that domestic oil furnaces are usually not more efficient than 70%, the equivalent fuel oil costs were calculated. The results are close to present fuel oil costs and below electric costs in most areas. Observing that the principal cost is debt service, which is inflation proof, it would appear that this would be an attractive investment even without the 40% credit up to $4,000 allowed on solar installations. The bottom two columns in Table 5 show the significant cost reduction resulting from the credit and indicate extremely attractive profitablility.

TABLE 5

SELECTED OPERATING COSTS
New Mexico Climate
Cost Case C
No Tax Credit

	With addition for distribution	Distribution equals furnace credit
Capital Cost	5770	4770
Debt Service	736	608
Maintenance	289	239
Pumping	50	50
Operating Cost ($/yr)	1075	897
Fuel oil equivalent $/Gal	0.94	0.78
With Tax Credit		
Capital	3462	2862
Debt Service	441	365
Maintenance	289	239
Pumping	50	50
Operating Cost ($/yr)	780	654
Fuel oil equivalent $/gal	0.68	0.57

A shallower, 500 GPM well, Case A, Figure 8, would reduce the investment per house by approximately another $1,000 and annual operating costs by $126. This is approaching natural gas in price and might be considered as an inflation hedge. Of course if Case B costs could be realized, the system could compete with gas now in good locations.

Calculations at the northern location indicate costs after the tax credit similar to those before the credit at the other location, so even here if fuel oil or electricity is the alternative, the investment is attractive.

CONCLUSIONS

The optimum aquifer storage-solar heater system is a complex function of weather, solar, aquifer and cost parameters which must be evaluated in each case. Except for very expensive wells, the heater cost predominates, and it can be more economical to drill more wells than to lose efficiency by operating the wells at peak load. For large heater areas to well flow volumes, it definitely pays to use some surface storage.

The gain in annual heat collection per unit of heater area over a conventional system will vary according to the design of both systems but it is probably in the range of three to five fold. Well costs must be added, but in many cases this will be offset by the saving in large, summer oriented heaters relative to the usual small, roof installed, winter oriented heaters. The cost of the distribution system, though large and highly variable, can easily be offset by the elimination of individual auxilary furnaces.

With existing federal tax credits, a solar heater-aquifer storage system can compete with oil under most circumstances and with gas under favorable circumstances. With a virtual freedon from inflation, it would appear to be a very attractive investment.

LITERATURE CITED

1. Davison R.R., W.B. Harris, and J.H. Martin Chemtech, 5, 736 (1975)

2. Davison, R.R., W.B. Harris, and D.L. Reddell, AIChE Symp.Ser., No. 198, p.34 (1980)

3. Ebeling, L., D.L. Reddell, S. Totengard, W.B. Harris, and R.R. Davison, "The Effect of System Size on the Practicality of Aquifer Storage" Proceedings of the International Solar Energy Soc., V. 1, p.269 (June 1979)

4. Harris, W.B., and R.R. Davison, U.S. patent 3,931,851 (Jan. 1976)

5. Meyer, C.F., "Large Scale Thermal Energy Storage for Cogeneration and Storage Systems" 5th Energy Technology Conference Exhibition, Washington D.C., (Feb. 1978)

6. Klien, S.A., Solar Energy, 17,79 (1975)

7. Hottel, H.C. and B.B. Woertz, Trans ASME, 64,91 (1942)

8. Bennett, I. Solar Energy, 9,145 (1965)

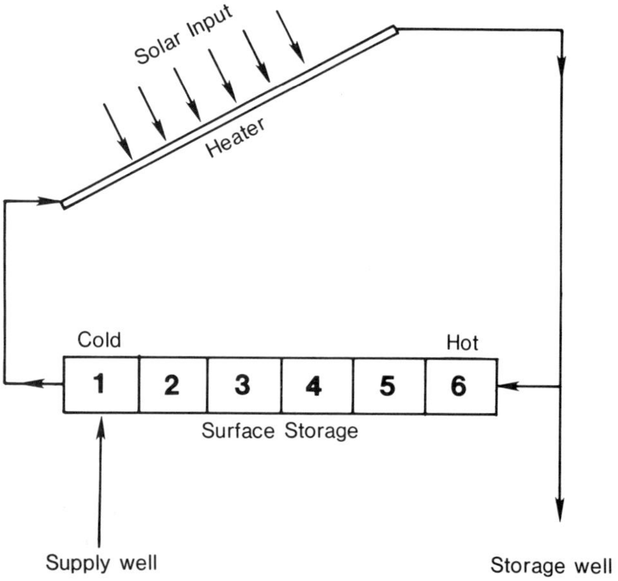

Figure 1. The simulated solar heater-surface storage arrangement.

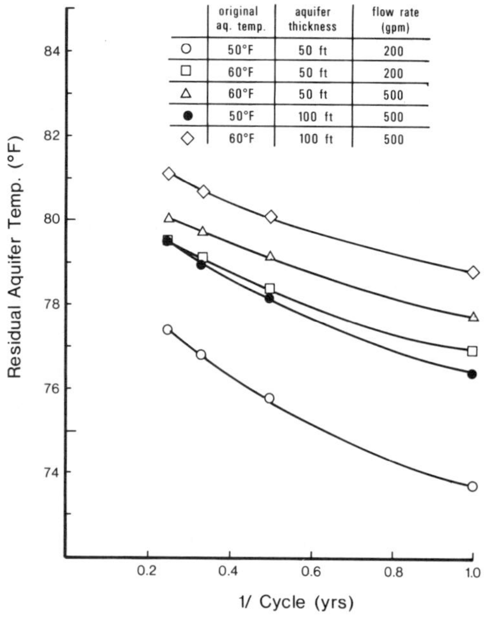

Figure 2. Water temperature from the supply well.

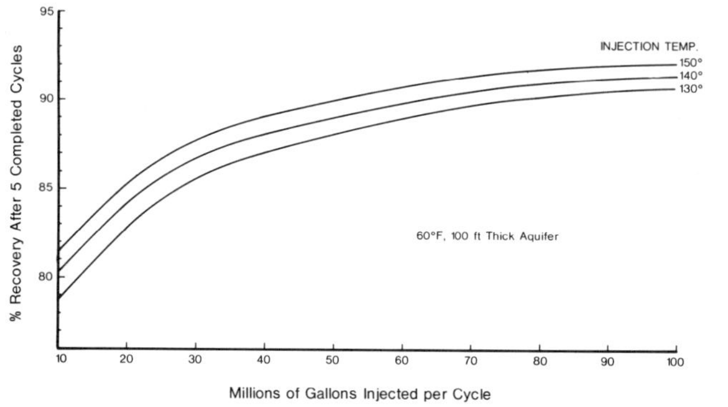

Figure 3. Recovery efficiency for a 100ft aquifer at warmer location.

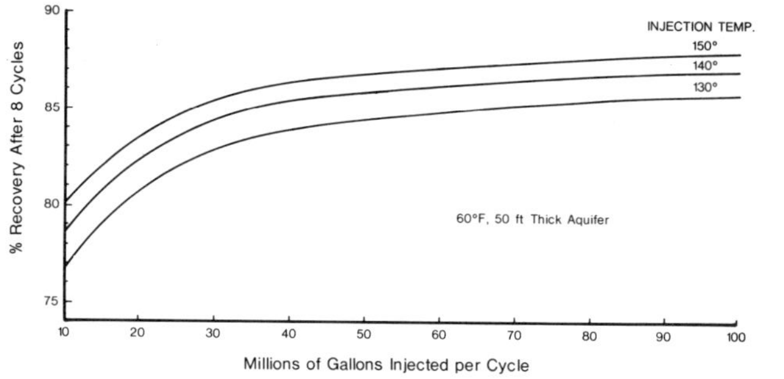

Figure 4. Recovery efficiency for a 50ft thick aquifer at warmer location.

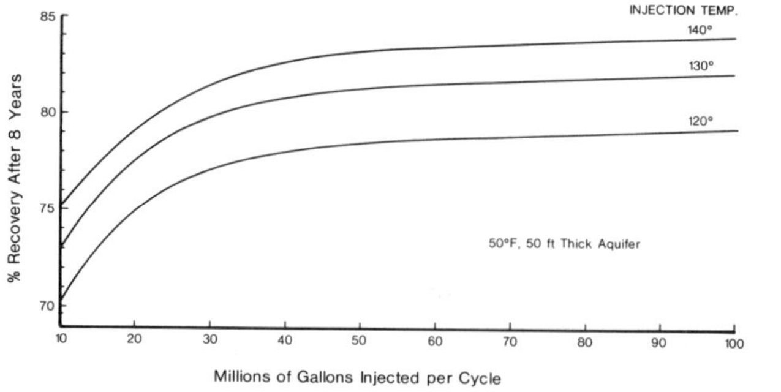

Figure 5. Recovery efficiency for a 50 ft thick aquifer at colder location.

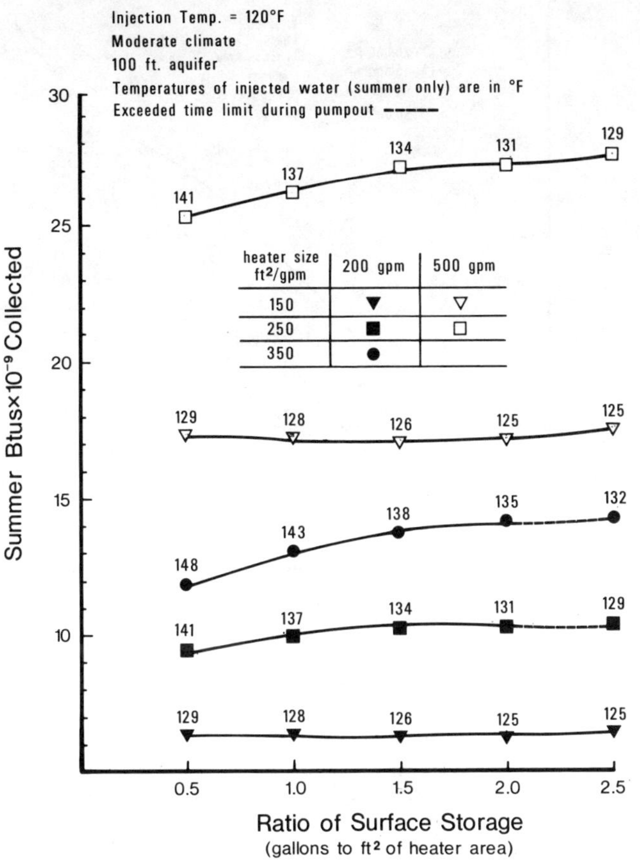

Figure 6. Summer heat collection.

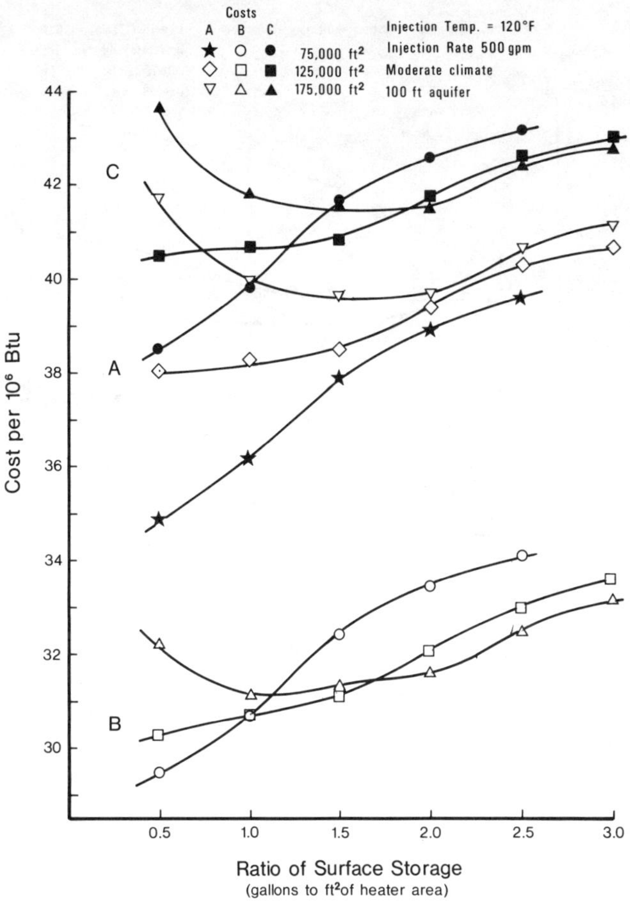

Figure 8. The effect of cost and heater area on optimum surface storage for 500 GPM wells.

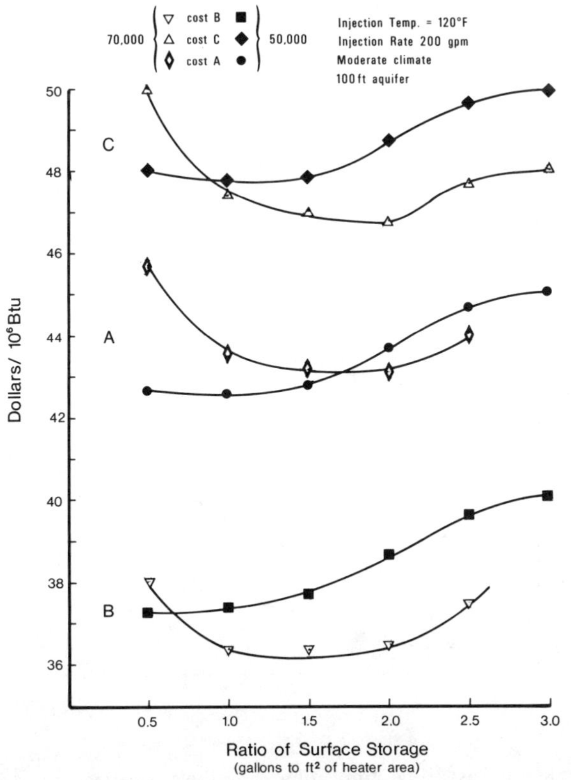

Figure 7. The effect of cost and heater area on optimum surface storage for 200 GPM wells.

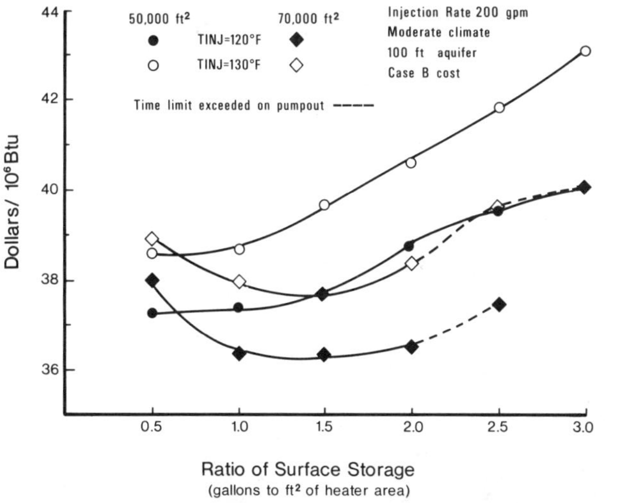

Figure 9. The effect of minimum injection temperature on cost and optimum surface storage.

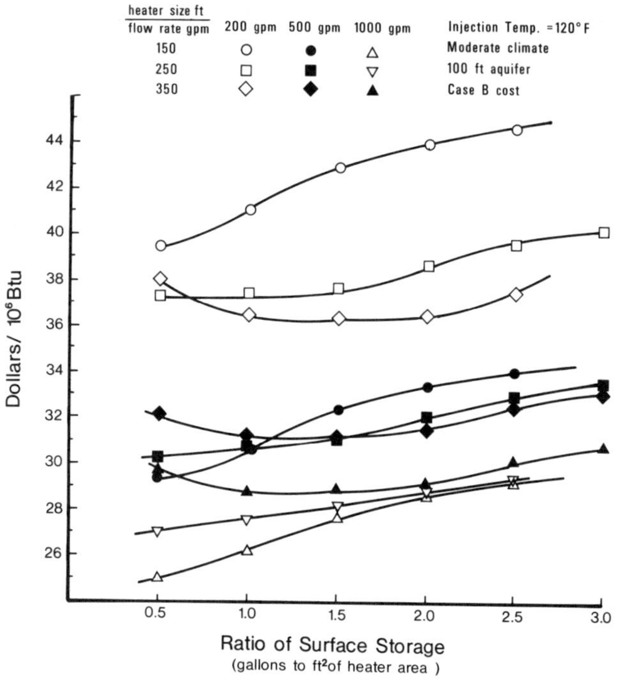

Figure 10. The effect of well capacity on cost and optimization.

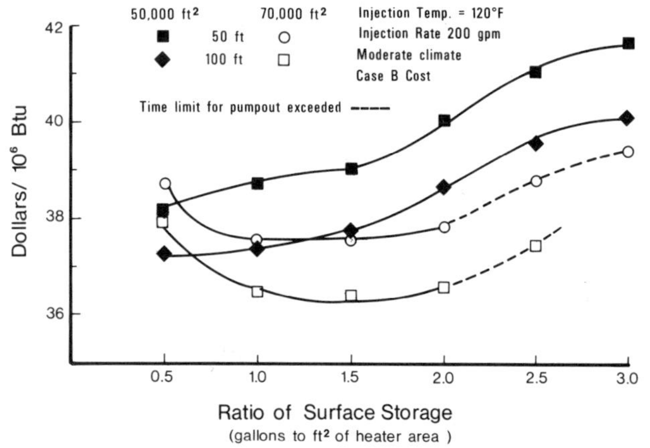

Figure 11. The effect of aquifer thickness on cost.

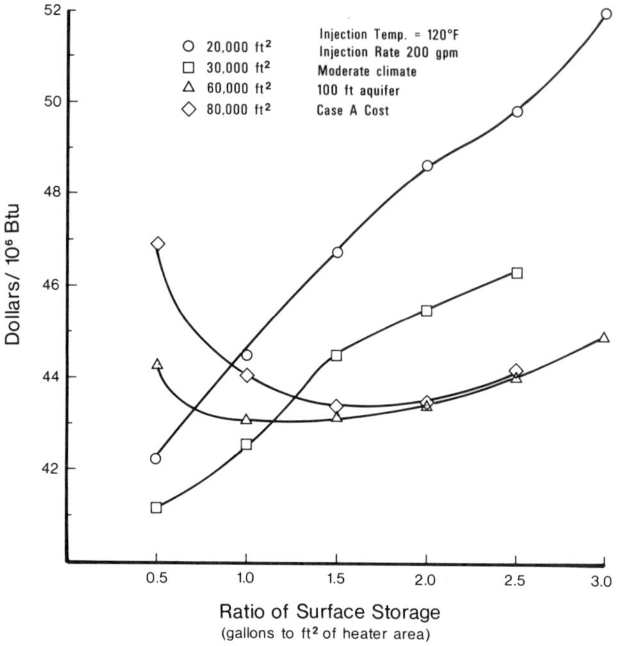

Figure 12. Variation of cost and optimum surface storage with heater area.

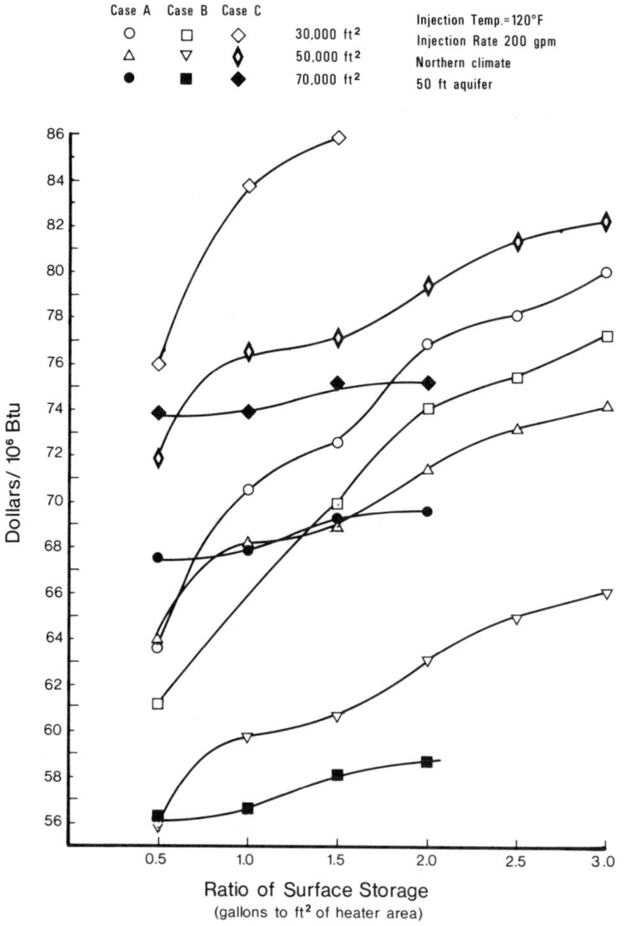

Figure 13. Cost and optimization at northern location.

HYDROCARBON OIL STORAGE FLUID MAINTENANCE: EXPERIMENT, ANALYSIS, AND ECONOMICS

D. G. BESHORE, P.E.
and
D. G. RAYNOR

Martin Marietta Aerospace
Denver, Colorado

Hydrocarbon-based oils at 316°C (600°F) are being considered by the U.S. Department of Energy for use as thermal energy storage and heat transport media in solar thermal power plants and solar industrial applications. An experimental and analytical research program has been conducted by Martin Marietta Aerospace under DOE contract EY-76-C-03-1110 and Sandia Laboratories-Livermore contract SAN 83-2754 to evaluate the thermal stability of Exxon's Caloria HT-43 oil at storage temperatures of 316°C (600°F). This paper highlights conclusions of this effort which have led to an understanding of the oil's thermal decomposition process, fluid replenishment requirements, and methods to reduce the oil degradation rate. Tests and analyses from this research program have shown that the thermal decomposition process results from a combination of both thermal cracking and autocatalysis. The results of our engineering scale tests clearly indicate that degradation reductions can be expected with reprocessing.

Preliminary costs have been developed for the processor unit based on capital, operating, and maintenance expenditures. A cost-to-savings of storage oil analysis was performed on an on-line reprocessing unit providing cost/savings ratios ranging from 0.21 to 0.50 for a processor installation at the solar central receiver 10 MWe Barstow pilot plant.

A NOVEL CHLOROPHYLL SOLAR CELL

A novel solar battery is being developed which uses chlorophyll a for the generation of a voltage. The battery consists of platinum foil electrode, onto which a mixture of chlorophyll a and lipoic acid is deposited, and a platinum current collector. With such a device, voltages greater than 0.35 volts can reproducibly generated. The dependence of the output of the cell as a function of chlorophyll levels and light intensity has been determined.

J. CHRISTOPHER LUDLOW

Department of Chemical Engineering
West Virginia University
Morgantown, West Virginia 26506

The scarcity of fossil fuels and their ultimate depletion has motivated researchers to investigate alternative sources of energy. One such source is solar energy. Of all the possible uses of solar energy, however, the conversion of sunlight into electricity may be the most desirable. The technology for controlling and using electrical energy is well established; its generation need not have a thermodynamic cycle as an intermediary, and direct production of electricity from sunlight would help conserve fossil fuels used to generate electricity.

Of the three possible effects that could be used to convert light into electricity, the photoelectric effect, the thermoelectric effect, and the thermionic effect, only the photoelectric effect has been developed to the point where devices may be purchased which employ this principle. The silicon solar cell which has been used so profitably in the space program is an example of such a device.

While this is theoretically a simple device, nowhere in nature is such a device found. Instead, nature's solar cell is a leaf, and the process by which part of the estimated 1.2×10^{24} J/year (1) of solar energy reaching the earth's surface is converted to electronic energy and then to chemical energy is called photosynthesis. While only roughly 1/2000 of this energy is ultimately converted to chemical energy by plants (2), the estimated yearly energy consumption of all man-made machines on earth is still only 3% (3) of this 6×10^{20} J/year figure.

Careful examination of the photosynthetic process indicates that the initial electronic excitation of chlorophyll occurring upon absorption of a photon maybe used to produce electricity instead of converting carbon dioxide and water into sugar and oxygen. To understand how a chlorophyll solar cell might be constructed, a knowledge of photosynthesis is necessary. The chlorophyll a molecule, the major photo active chemical within a leaf, pictured in Figure 1 can be thought of as being organized into two basic structures within the photosynthetic apparatus. The first structure, consisting of 200-300 molecules, is termed antenna chlorophyll and serves to absorb the light energy and direct it to the second structure. This second structure is an energy trap and is thought to be made up of protein and two chlorophyll molecules (4). It is the chlorophyll trap which actually initiates the oxidation-reduction chemistry associated with photosynthesis. The photosynthetic process can be thought of as occurring in a manner similar to that pictured in Figure 2.

0065-8812-81-4476-0210-$2.00

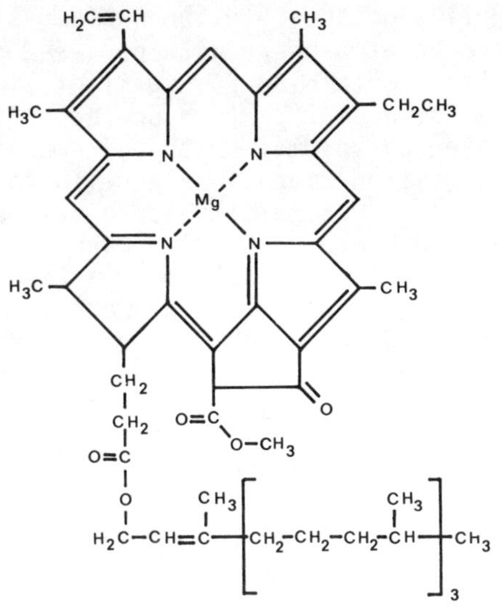

FIGURE 1

CHLOROPHYLL a MOLECULE

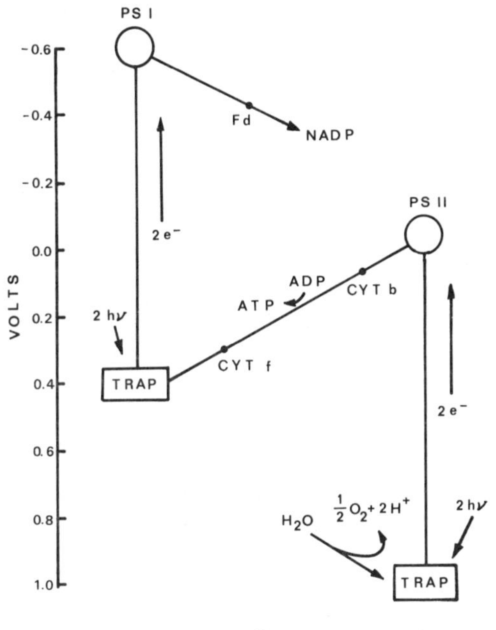

FIGURE 2

ENERGY DIAGRAM FOR PHOTOSYNTHESIS

Upon absorption of a photon of energy and subsequent trapping, the redox potential of the chlorophyll trap changes from +0.4 to approximately -0.6 volts. If there exists adjacent to the excited traps, a molecule whose redox potential is slightly more than -0.6 volts, -0.43 (as in the case of ferridoxin), then the excited trap can reduce that molecule. This reduction is nothing more than the transfer of an electron from the trap to the accepting molecule. The "hole" left in the trap by the lost electron is filled by an electron from an appropriate electron donor. The unexcited trap is then ready to undergo another cycle. The molecule which has accepted the electron can then reduce other molecular species whose redox potential is closer to zero than its own. It is during this transfer of the electron from molecule to molecule that much of the electrochemical work of photosynthesis occurs.

THE CHLOROPHYLL CELL

If the antenna chlorophyll molecules and their associated energy traps of one of the photosystems, say Photosystem I (PSI), were to be removed from the leaf and placed in an environment which had the necessary electron acceptors, donors, and electrodes, then the electron which is transferred from the energy traps could be channeled outside the system in the form of electricity. The result would be a battery, solar powered, containing chlorophyll. A schematic diagram of such a cell is pictured in Figure 3. The chlorophyll antenna and traps are placed on a metallic support which is bathed in an electron acceptor solution. When light strikes the chlorophyll, the excited trap reduces the acceptor. The acquired electron is transferred through the solution to a metallic current collector where the electron is given up. The electron hole left within the chlorophyll is filled by an electron from the metallic support.

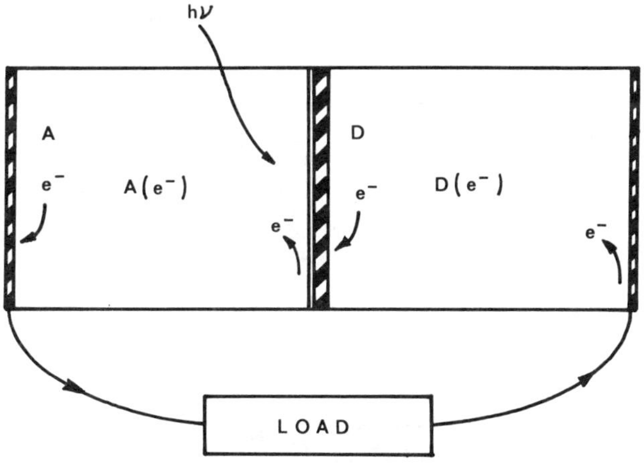

FIGURE 3

SCHEMATIC DIAGRAM

This process, however, leaves a hole in the metal, which is filled by an electron from an electron donor. The donor maintains electrical neutrality by accepting an electron from another metallic current collector. The net result is an electron flow, or electricity, induced by sunlight.

The construction of such a device is not a new idea. Douglas and Albrecht (5) investigated photo produced electrons from chlorophyll by suspending chlorophyll a in 3-methylpentane, placing this suspension in a conductivity cell and applying a potential of 100 volts across the cell. Photo-induced potential changes were then monitored at room temperature and at 77°K.

One of the more typical solar cells that have been proposed previously consists of two solutions, each in contact with an electrode, separated by a membrane containing chlorophyll. Such a configuration, employing tetramethylphenylenediamine in one compartment and sodium ascorbate in the other, was investigated by Katz, Wasielewski, and Janson (6). Their cell contained chlorophyll impregnated in a plastic membrane separating the two solutions. The cell was capable of generating a potential of about 0.4 volts. A similar cell was investigated by S. J. Valentz (7) using substituted p-benzoquinones and chlorophyll suspended in a bilayer lipid membrane.

Another type of chlorophyll solar cell configuration consists of, once again, two compartments containing two solutions. However, in this case, the solutions are separated by a fritted glass plug and the chlorophyll is deposited upon one or both of the electrodes. Such a cell has been investigated by F. K. Fong (8) and by F. Takahashi (9) with potentials on the order of 0.25 volts being produced.

The chlorophyll cell used in this investigation is somewhat simpler than those investigated before. A schematic diagram of the cell is pictured in Figure 4. The cell was constructed of three major pieces: 1) a back consisting of a 10 cm x 8.25 cm x 0.127 cm microscope slide onto which a 5 cm x 5 cm x 0.004 cm piece of platinum foil was cemented; 2) a front piece of glass onto which a set of platinum fingers (acting as a current collector) was cemented; and 3) a 8.25 cm x 8.25 cm x 0.381 cm Teflon spacer to separate the front and the back. Open circuit photopotentials produced by light from a 150 watt lamp (with the infrared filtered out) were measured with a high impedance amplifier and digital voltmeter. See Figure 5.

The chlorophyll used in these experiments was extracted from spinach, purified, and then applied to the platinum support in the form of an octane-water emulsion. The chlorophyll n-octane solution was emulsified in a nitrogen atmosphere with degassed distilled water containing 0.5 mg/ml of liporic acid at a ratio of 7 parts hydrocarbon to 1 part water by volume. With the emulsion made, the desired amount was applied to the platinum electrode and allowed to dry in a nitrogen atmosphere. Typical amounts of chlorophyll left on the platinum after evaporation of the water and octane were 2.6 x 10^{-4} gr/cm^2. With the chlorophyll dried, the cell is assembled using a 0.5M NaCl solution in the cavity formed by the Teflon spacer. The cell was allowed to equilibrate in the dark for two hours before being illuminated. After equilibration, three hours of filtered light were used to investigate the photoresponse. The intensity of the illumination was measured at 590 W/m^2. Typical voltage response versus time curves are represented in Figure 6. As can be seen, when the lamp is turned on there is an initial jump in voltage of about 0.15 volts. After this initial rise, the voltage increases are less rapid until after about 3 hours, when the maximum voltage of about 0.35 volts is reached.

The effects of light intensity on produced voltage was also investigated. These results are displayed in Figure 7. Here light intensities ranged from 590 W/m^2 as before to 10 W/m^2. In each case, the cell's make up was the same as in the initial set of experiments. The obvious dependence of the cell's output on light intensity indicates that not only does the voltage take longer to reach a maximum at low intensities but also this maximum may be less.

Perhaps the ultimate test of a solar cell is its performance in direct sunlight. Such a test was performed with the results shown in Figure 8. Here the photo potential and light intensity is graphed as a function of time. The initial voltage response upon illumination can be seen to vary similarly to that obtained for the artificially lit cell. While the light intensity varies throughout the experiments, corresponding changes in output in the cell, however, are not apparent. Instead, only very slow voltage fluctuations are noticed. Because of the lack of close correspondence in frequencies of fluctuations, the changes in output are not thought to be solely intensity induced. In fact, the lack of intensity dependent fluctuations indicates saturation of the process involved.

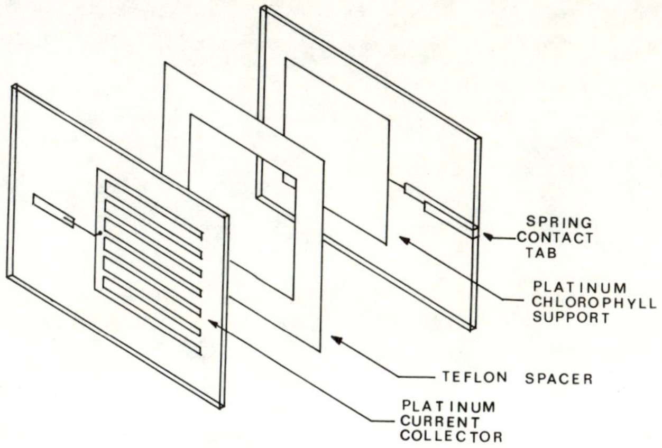

FIGURE 4

DIAGRAM OF CELL

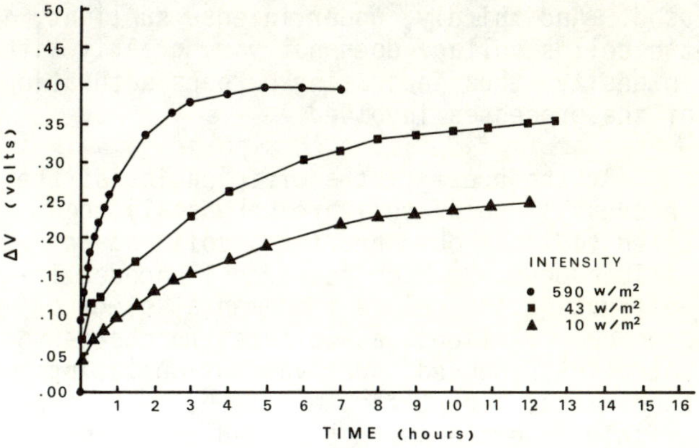

FIGURE 7

PLOT OF VOLTAGE vs TIME
FOR DIFFERENT LIGHT INTENSITIES

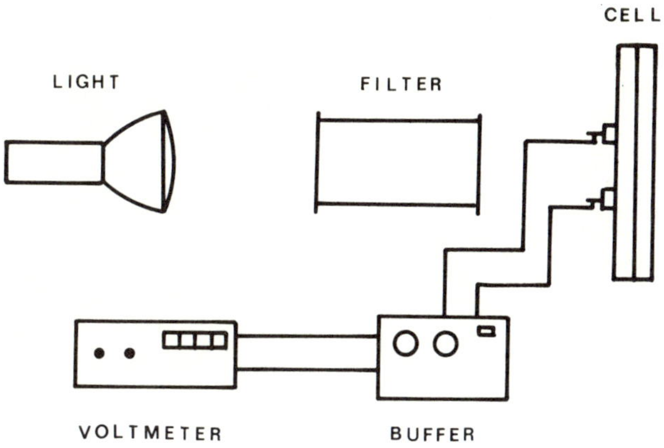

FIGURE 5

SCHEMATIC DIAGRAM
OF APPARATUS

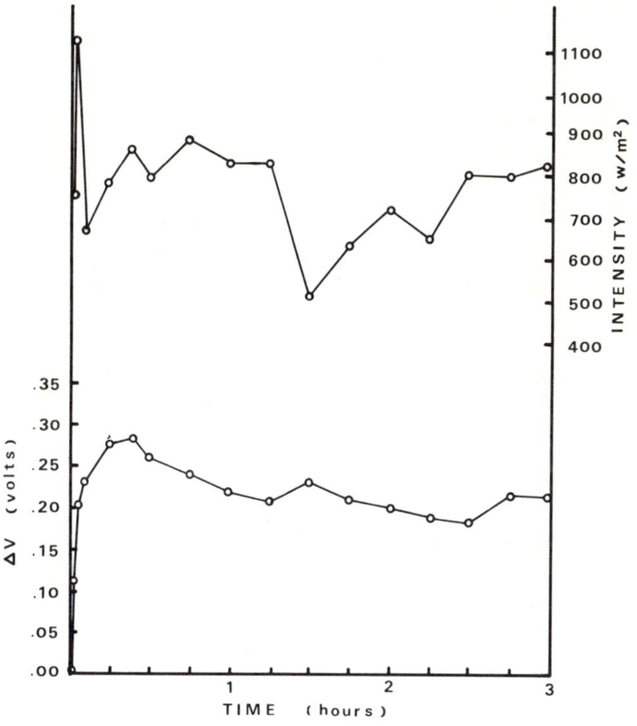

FIGURE 8

PLOT OF VOLTAGE & INTENSITY vs TIME

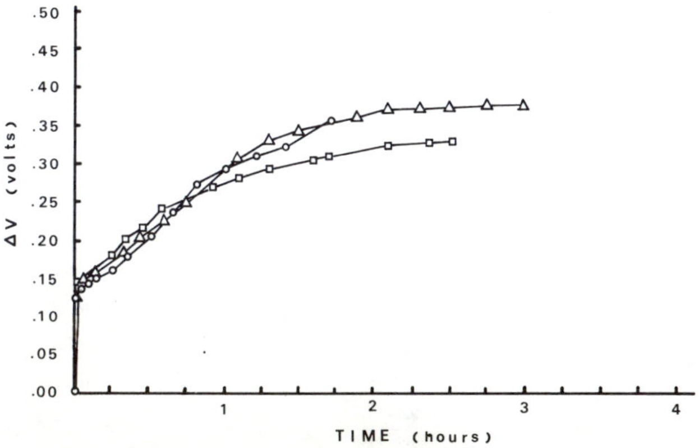

FIGURE 6

PLOT OF VOLTAGE vs TIME

CONCLUSIONS

The first and perhaps most important conclusion which can be drawn is that it appears to be possible to remove parts of the photosynthetic process from a green leaf and use it to produce electricity. Secondly, the voltage versus time response of the cell produced using chlorophyll a has been shown to be dependent upon the intensity of light used. And thirdly, under intense sunlight, the cell's voltage does not vary greatly with intensity, thus indicating perhaps saturation of the processes involved.

At the present, the efficiencies of the chlorophyll solar cell are very small compared to those obtained for a solid state silicon device. Much research is needed investigating the cell's fundamental electrochemical reactions, as well as the photopotential produced under various operating conditions. The ultimate goal being a more efficient device which supposedly will mimic the photosynthetic process developed over millions of years of evolution.

LITERATURE CITED

1. Stokes, Seager, and Carpener. Energy--From Source to Use, Glenview, Illinois, Scott Foresman & Co., 1975.

2. Lehninger, A. L., Bioenergetics, Menlo Park, Calif.: Wa. A. Benjamin, 1971.

3. Ibid.

4. Fenton, J. M., M. J. Pellin, Govindjee, and K. J. Kaufman, FEBS Letters, 100:1, 1-4, 1979.

5. Douglas, F. and A. C. Albrecht, Chem. Phys. Lett., 14:150-54, 1972.

6. Chemical Week, 9 February 1977, pp. 41-42.

7. Valenty, S. J., An Organic Solar Cell, Technical Information Series, Report No. 73CRD091, Schenectady, NY: General Electric Co., Corporate Research and Development, March 1973.

8. Fong, F. K. and N. Winograd, J. Am. Chem. Soc., 98:8, 2287, 1978.

9. Takahashi, F., Biochim Biophys Acta, 430:3, 490-500, 1976.

THEORETICAL STUDY OF ABSORBED SOLAR ENERGY IN MULTI-LAYER ABSORBER COATINGS FOR RECEIVERS OF SOLAR CONCENTRATORS: PART I. RADIATION TRANSFER ANALYSIS

The use of solar energy to produce electricity is under development. High temperature fluid is heated by flowing through blackened receivers placed at the focus of concentrating collectors. This fluid is used to drive a conventional turbine working on Brayton cycle. The receivers should have high absorptance for short wave radiation and low emissivity at the infra-red radiation. Several receiver coatings have been suggested in the literature. A semiconductor coating has the advantage of being transparent to long wave radiation but it has the disadvantage of having high reflectance. The present work analyzes a multi-layer coating composed of a thin semi-conductor layer covered by two very thin absorbing layers. The coating lies on a thin layer of a low-emitting silver substrate. The multiple reflections and multiple absorptions are considered in deriving a mathematical model. The effect of absorption coefficients and thickness of the three layers, as well as the reflection coefficient of the semiconductor layer, on the solar energy absorbed within the different layers of the coating is studied. A discussion is made on how to apply the results of this model to evaluate the design of the receivers.

I. S. TAHA
M. A. DARWISH
and
M. M. ELSAYED

Mechanical Engineering Department
King Abdulaziz University
P.O. Box 9027
Jeddah, Saudi Arabia

INTRODUCTION

Utilization of solar energy for power production is developing slowly as a result of low overall efficiency encountered in the process. The thermal efficiency of the power cycle increases as the high temperature of the medium collecting the solar energy increases. Meanwhile, the collection efficiency decreases due to increased losses to the surroundings. Highly concentrating solar collectors, with concentration of 40:1 to 2000:1 , having central or distributed receivers are designed to increase the collecting temperature to values ranging between 300 and 800 °C (1). Research is thus directed towards increasing the collecting efficiency of receivers of highly concentrating collectors.

The receivers should absorb the maximum possible shortwave radiation in the solar spectrum and at the same time should emit the lowest possible long-wave radiation in the infrared spectrum. Also, other thermal losses by conduction and convection to the surroundings should be minimized. Melamed and Kaplan (1) made an extensive survey for different coatings which meet some of the previous requirements. The survey showed the use of a semiconductor layer in some coatings where it is deposited on a micro layer of a low emitting material like silver. The semiconductor has the advantages of absorbing short-wave solar radiation beside being transparent to long-wave emitted radiation. However, it has the disadvantage of high reflectance to solar radiation (about 35%). The latter disadvantage can be overcome by covering the semiconductor by micro layers of absorbing materials which recover most of the reflected radiation from the semiconductor without jeopardizing the emitting characteristics of the coating.

To select the materials of the coating and their thickness, the process of solar interaction with the receivers' coating components has to be studied. Unfortunately, there are very few data on the materials optical properties which are necessary to study the process completely. However, the authors feel that the theoretical study presented here may promote research in the area of measuring the optical properties of different materials to complete this study.

This study attempts to analize qualitatively a model describing the

0065-8812-81-4012-0210-$2.00

short-wave radiation interaction with a coating composed of a micro layer of low-emitting substrate covered by a semiconductor layer and two absorbing micro layers. The use of two top layers permits selection of layers having different optical characteristics and different thickness. It is thought that one of the layers has to be very absorbing (which is expected to be more or less opaque to long-wave radiation) and hence relatively small thickness can be used without sacrificing the long-wave transparency of the semiconductor. The other layer may be relatively thick and having relatively low absorptance for solar radiation.

ANALYSIS

Figure 1 shows a cross section in a multi-layer coating composed of four layers. The thickness of the first two layers and the fourth oneare of smaller order of magnitude than the third layer (semiconductor such as amorphous silicon). The materials of the first three layers are assumed to be diffuse (i.e. they diffuse the incident beam radiation uniformly).

The fourth layer is an opaque metal with very low emissivity (such as silver).

The top two layers are assumed to be totally absorbing, non-reflecting. Multiple absorptions take place as solar flux penetrates them. As solar flux penetrates the semiconductor layer, multiple absorptions and multiple reflections occur. The radiant flux reaching the fourth layer is partially absorbed at the surface and the rest is reflected back. For any infinitismal volume of thickness dX at location X from the top surface there are two radiation fluxes opposite to each other: the incoming flux Y and the outgoing flux Z (see Figure 2). The incoming flux Y is the sum of the transmitted flux due to the incident radiation plus the reflected portion of the outgoing flux. The outgoing flux Z is the sum of the transmitted flux from the semiconductor and its substrate plus the reflected portion of the incoming flux.

For diffuse material the spectral absorption and reflection coefficients are respectively defined as follows (see Figure 1).

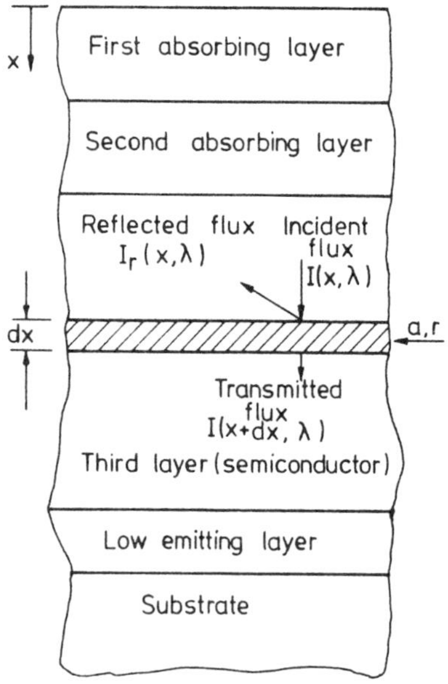

Figure 1 Cross section in the Receiver's Coating.

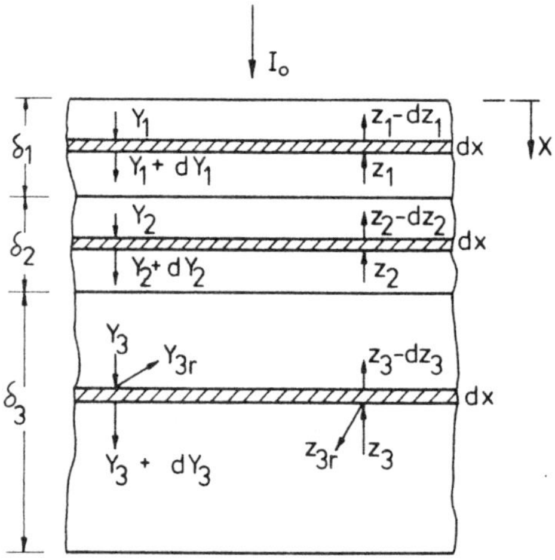

Figure 2 Cross section in the Top Two Layers and the Semiconductor Layer Showing the Elemental Volumes

$$a_\lambda = \lim_{dX \to 0} \frac{I_{\lambda,a}(X)}{I_\lambda(X).dX} \quad (1)$$

$$r_\lambda = \lim_{dX \to 0} \frac{I_\lambda(X) - I_\lambda(X+dX) - I_{\lambda,a}(X)}{I_\lambda(X).dX} \quad (2)$$

where

a_λ =monochromatic absorption coefficient, m^{-1}

r_λ =monochromatic reflection coefficient, m^{-1}

$I_\lambda(X)$ =incident monochromatic radiation flux on the elemental volume, $W/m^2\ \mu m$

$I_\lambda(X+dX)$ =transmitted monochromatic radiation flux from the elemental volume, $W/m^2\ \mu m$

$I_{\lambda,a}(X)$ =absorbed monochromatic radiation flux by the elemental volume, $W/m^2\ \mu m$

It is to be noted that all the optical quantities hereafter are monochromatic and the subscript λ will be dropped to simplify notation.

Figure 2 shows elemental volumes and boundary conditions for the top two layers and the semiconductor layer of the coating irradiated by a monochromatic flux I_o. The subscripts 1, 2, and 3 for Y and Z shown in the Figure correspond to the three layers respectively. The second subscript r in the flux symbols denotes reflected portions of the penetrating flux. Provided that the top two layers are nonreflecting, the mathematical model is presented by the following differential equations and boundary conditions.

Differential Equations:

For $0 < X < \delta_1$

$$dY = -a_1\, dX\, Y_1 \quad (3)$$

$$dZ_1 = a_1\, dX\, Z_1 \quad (4)$$

For $\delta_1 < X < \delta_1 + \delta_2$

$$dY_2 = -a_2\, dX\, Y_2 \quad (5)$$

$$dZ_2 = a_2\, dX\, Z_2 \quad (6)$$

For $\delta_1 + \delta_2 < X < \delta_1 + \delta_2 + \delta_3$

$$dY_3 = -(a_3 + r_3) dX\, Y_3 + r_3\, dX Z_3 \quad (7)$$

$$dZ_3 = (a_3 + r_3) dX\, Z_3 - r_3 dX\, Y_3 \quad (8)$$

Boundary Conditions:

at $X=0$ $\quad Y_1 = I_o \quad (9)$

at $X=\delta_1$ $\quad Y_1 = Y_2 \quad (10)$

$Z_1 = Z_2 \quad (11)$

at $X=\delta_1+\delta_2$ $\quad Y_3 = (1-\rho_3) Y_2 + \rho_3 Z_3 \quad (12)$

$Z_2 = (1-\rho_3) Z_3 + \rho_3 Y_2 \quad (13)$

at $X=\delta_1+\delta_2+\delta_3$ $\quad Z_3 = \rho_4\, Y_3 \quad (14)$

where ρ_3 and ρ_4 are the monochromatic reflectivities of the semiconductor surface and the semiconductor-substrate interface respectively.

Equations(3) to (6)are first order, first degree homogeneous differential equations. Their solutions have exponential forms in the independent variable X . Equations (7) and (8) are two simultaneous first order, first degree equations which can be reduced, by simple manipulation, to the following second order, first degree equation for Y_3

$$\frac{d^2 Y_3}{d\, X^2} - [\,(a_3 + r_3)^2 - r_3^2\,]\, Y_3 = 0 \quad (15)$$

Equation (15) is solved for Y_3. The solution is substituted into Equation (7) to obtain Z_3.

The general solution has six unknowns which are determined by the boundary conditions given by Equations (9) through (14). The final solution is obtained as follows:

$$Y_1 = I_o\, e^{-a_1 X} \quad (16)$$

$$Z_1 = (1-\rho_3) \frac{(a_3 + r_3 + \beta)}{r_3} A_3\, e^{-a_1(2\delta_1 - X) - 2a_2\delta_2 - 2\beta\delta_3} + (1-\rho_3) \frac{(a_3 + r_3 - \beta)}{r_3} A_4$$

$$e^{-a_1(2\delta_1-X)-2a_2\delta_2}$$

$$+\rho_3 I_o e^{-a_1(2\delta_1-X)-2a_2\delta_2} \qquad (17)$$

$$Y_2 = I_o e^{-a_1\delta_1-a_2(X-\delta_1)} \qquad (18)$$

$$Z_2 = (1-\rho_3)\frac{(a_3+r_3+\beta)}{r_3} A_3 \text{Exp}[-a_1\delta_1 - a_2(2\delta_2+\delta_1-X)-2\beta\delta_3]$$

$$+(1-\rho_3)\frac{(a_3+r_3-\beta)}{r_3} A_4 \text{ Exp}[-a_1\delta_1 - a_2(\delta_1+2\delta_2-X)]$$

$$+\rho_3 I_o \text{ Exp}[-(a_1+a_2)\delta_1-a_2(2\delta_2-X)] \qquad (19)$$

$$Y_3 = A_3 \text{ Exp}[-(a_1+\beta)\delta_1-(a_2+\beta)\delta_2 - \beta(2\delta_3-X)]$$

$$+ A_4 \text{ Exp}[(\beta-a_1)\delta_1+(\beta-a_2)\delta_2-\beta X] \qquad (20)$$

$$Z_3 = \frac{(a_3+r_3+\beta)}{r_3} A_3 \text{ Exp}[-(a_1+\beta)\delta_1 -(a_2+\beta)\delta_2-\beta(2\delta_3-X)]$$

$$+\frac{(a_3+r_3-\beta)}{r_3} A_4 \text{ Exp}[(\beta-a_1)\delta_1 + (\beta-a_2)\delta_2-\beta X] \qquad (21)$$

where

$$\beta = \sqrt{(a_3 + r_3)^2 - r_3^2} \qquad (22)$$

$$A_3 = \frac{(1-\rho_3) I_o}{(d+b)} \qquad (23)$$

$$A_4 = c A_3 \qquad (24)$$

$$d = [1-\frac{\rho_3(a_3+r_3+\beta)}{r_3}]\text{Exp}(-2\beta\delta_3) \qquad (25)$$

$$b = \frac{[(a_3+r_3+\beta)/r_3-\rho_4]\ [1-\rho_3(a_3+r_3-\beta)/r_3]}{[\rho_4 - (a_3+r_3-\beta)/r_3]} \qquad (26)$$

$$c = \frac{b}{[1-\rho_3(a_3+r_3-\beta)/r_3]}$$

The above solution can be used to predict the total amount and the distribution of the energy absorbed within the three layers per unit volume per unit wave length as follows

$$q_1''' = a_1 (Y_1 + Z_1) \qquad (28)$$

$$q_2''' = a_2 (Y_2 + Z_2) \qquad (29)$$

$$q_3''' = a_3 (Y_3 + Z_3) \qquad (30)$$

Since the incident flux and the optical properties are functions of wave length, the radiation fluxes Y and Z and the absorbed energy per unit volume q''' can be integrated over the spectrum of the incident flux to obtain the corresponding total quantities.

The spectral short-wave flux reflected from the coating is equal to the penetrating flux Z_1 calculated at the top surface of the coating. The spectral reflectance of the coating R_m is defined as the latter value divided by the spectral incident flux. The total or albedo reflected flux from the coating is the integrated value of the spectral reflected flux for the whole solar spectrum. The latter quantity when divided by the integrated solar flux results in the total or albedo reflectance of the coating.

The radiant flux absorbed at the semiconductor-substrate interface is equal to $(1-\rho_4)\, Y_3 \big|_{X=\delta_1+\delta_2+\delta_3}$. This quantity together with the energy absorbed within the coating are needed for the analysis of heat transfer in the coating.

ANALYTICAL RESULTS AND DISCUSSION

Data of monochromatic optical properties of prospect materials which can be used for an absorbing coating are generally lacking. A prospect material for the semiconductor is amorphous silicon whereas metal oxides are prospect materials for the other absorbing layers. Data of monochromatic absorption coefficient of amorphous silicon for part of the solar spectrum is reported by Reference (1). Calculations are carried out at a wave length of 0.5 μm at which monochromatic solar power is maximum (assumed to be 1293 W/m^2 μm). At this wavelength the absorption coefficient of amorphous silicon is found by extrapolation from Reference (1), to be 2×10^7 m^{-1} . The reflection coefficient of amorphous silicon is assumed to be 1×10^{-6} m^{-1} at 0.5 μm. Its upper surface reflectivity ρ_3 is assumed to be 0.33, whereas the reflectivity at the interface between silicon and the low emitting sublayer ρ_4 is assumed to be 0.97. The thickness of amorphous silicon δ_3 is given the value of 1.5 μm to suit selective coatings applications.

The other variables, which are the thickness and the absorption coefficients of the top two layers, are given a range of values to discuss their effect on the energy absorption and the energy reflected from the coating. Figures 3 and 4 show a semilogarithmic plot of the monochromatic energy absorbed per unit time per unit volume within the three diathermanous layers of the coating against the position from the top layer surface at different values of these variables. For both Figures, the thickness of the top layer δ_1 is fixed at a value of 0.1 μm. The absorption coefficient of the top layer a_1 is 1×10^5 m^{-1} for Figure 3 whereas its value is 1×10^6 m^{-1} for Figure 4. The thickness of the second top layer δ_2 for both Figures has a range of values from 0.1 to 1 μm while its absorption coefficient varies from 10^5 to 5×10^7 m^{-1}. Values of spectral reflectance of the coating R_m are calculated and given for each group of variables in Figures 3 and 4. The reflectance R_m is also shown in Figure 5 as a function of a_2 for different values of a_1 and δ_2 .

The calculations performed for micro-multilayer coatings have to be discussed only with regard to their field of application; namely, for selective coatings. In this regard, the interest is only to what percentage is absorbed within the coating which can be detected directly from

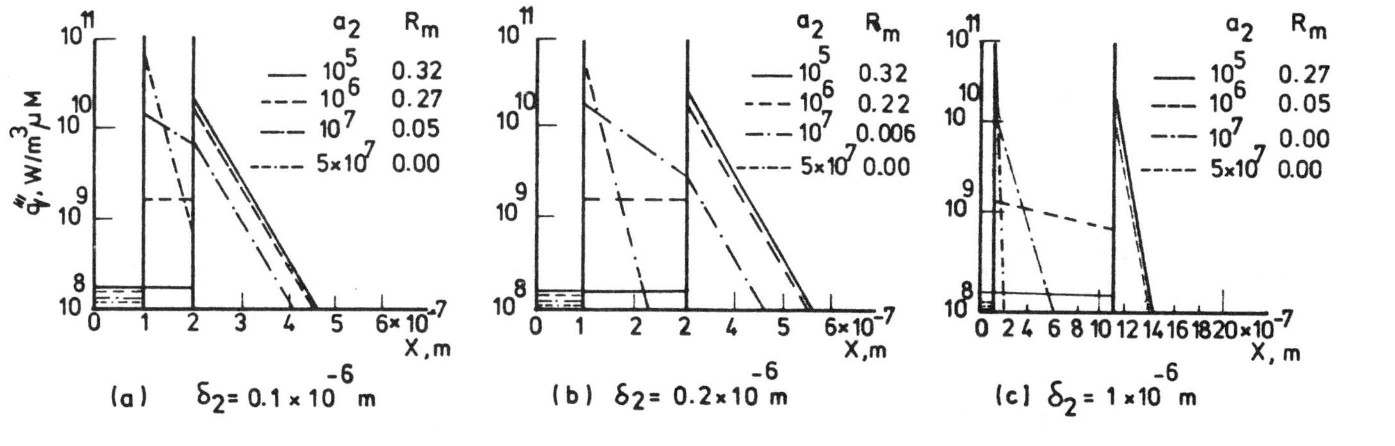

Figure 3 Effect of Absorption Coefficient a_2 on Energy Absorbed per unit Volume q''' ($a_1 = 1 \times 10^5$ m^{-1} , $\delta_1 = 1 \times 10^{-7}$ m)

reflectance values of the coating R_m. It can be detected directly from Figure 5 that a combination of relatively high values of a_1 , a_2 , and δ_2,within the range of values calculated, can lead to zero reflectance values. However, it is expected that the long-wave emittance is affected by both the thickness and absorption coefficients of the layer (or in other words the materials) of the top two layers. It is expected that a material of high optical thickness ($a\,\delta$) will be more opaque to long-wave emittance of the low-emitting substrate. A compromise between the energy gain by absorption and long-wave radiation loss from the coating should be made.

The way by which energy is absorbed for other applications of the model, when the layers' thickness are relatively large, is very important too. In this case, the temperature distribution across the layers is important and affects the conductive, convective, and radiative modes of heat loss from the coating. To reduce these losses, the variables should be chosen in accordance with both total energy absorbed and its being absorbed more in and near to the semiconductor layer. Figures 3 and 4 show that more energy is absorbed near the outer surface for higher values of a_1 , a_2 and δ_2 which means lower useful energy. Again, a compromise should be judged when selecting the parameters of the coating.

The discussion would have been more valuable if done on the total not the monochromatic quantities. The lack of data was discussed earlier. Blickensderfer et. al (2) presented methods of spectral evaluation of the optical properties of micro layers of different diathermanous materials. Data presented for the spectral reflectance for different thickness of the layers can be used to determine the spectral optical properties. This can be done by substituting the data into the solution of the mathematical model of this analysis or similar models constructed by applying the same principle. This results into some equations in the unknown optical properties. The resulting equations can be solved numerically for the required optical properties.

CONCLUSIONS

The solar radiation interaction with a coating composed of two layers of absorbing materials and a semiconductor layer is studied. The mathematical model presented predicts the spectral solar energy absorbed within the different layers of the coating. The discussion of the results of the model, using limited data, suggests that this type of coating can be used satisfactorily to decrease the energy losses if properly designed. Complete analysis of the problem necessitates developing techniques for measuring optical properties of different absorbing materials.

NOTATION

- a_i monochromatic absorption coefficient of the i-th layer, m^{-1}
- I_a monochromatic absorbed radiation flux, $W/m^2\ \mu m$
- I_r monochromatic reflected radiation flux, $W/m^2\ \mu m$
- q_i''' monochromatic absorbed radiation per unit volume at any depth in the i-th layer, $W/m^3\ \mu m$
- R_m monochromatic reflectance of the coating
- r_3 monochromatic reflection coefficient of the third layer, m^{-1}
- X depth from the top surface at any location in the coating, m
- Y_i monochromatic radiation flux penetrating the i-th layer in the direction of X, $W/m^2\ \mu m$
- Z_1 monochromatic radiation flux penetrating the i-th layer in the negative direction of X, $W/m^2\ \mu m$
- δ_i thickness of the i-th layer, m
- ρ_3 monochromatic reflectivity at the top surface of the semiconductor layer
- ρ_4 monochromatic reflectivity at the interface surface between the semiconductor layer and the substrate

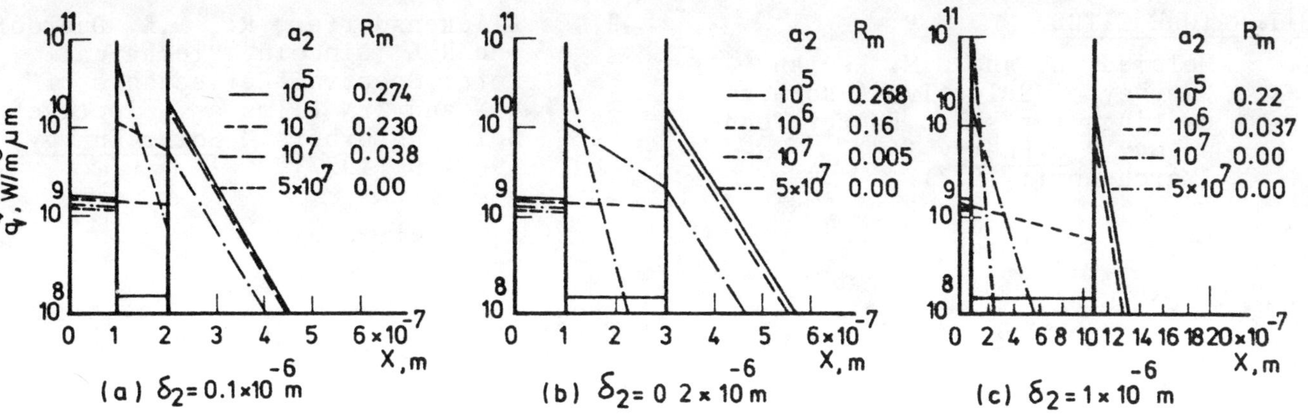

Figure 4 Effect of Absorption Coefficient a_2 on Energy Absorbed per Unit volume q''' ($a = 1 \times 10^{6}\ m^{-1}$, $\delta_1 = 1 \times 10^{-7}$ m)

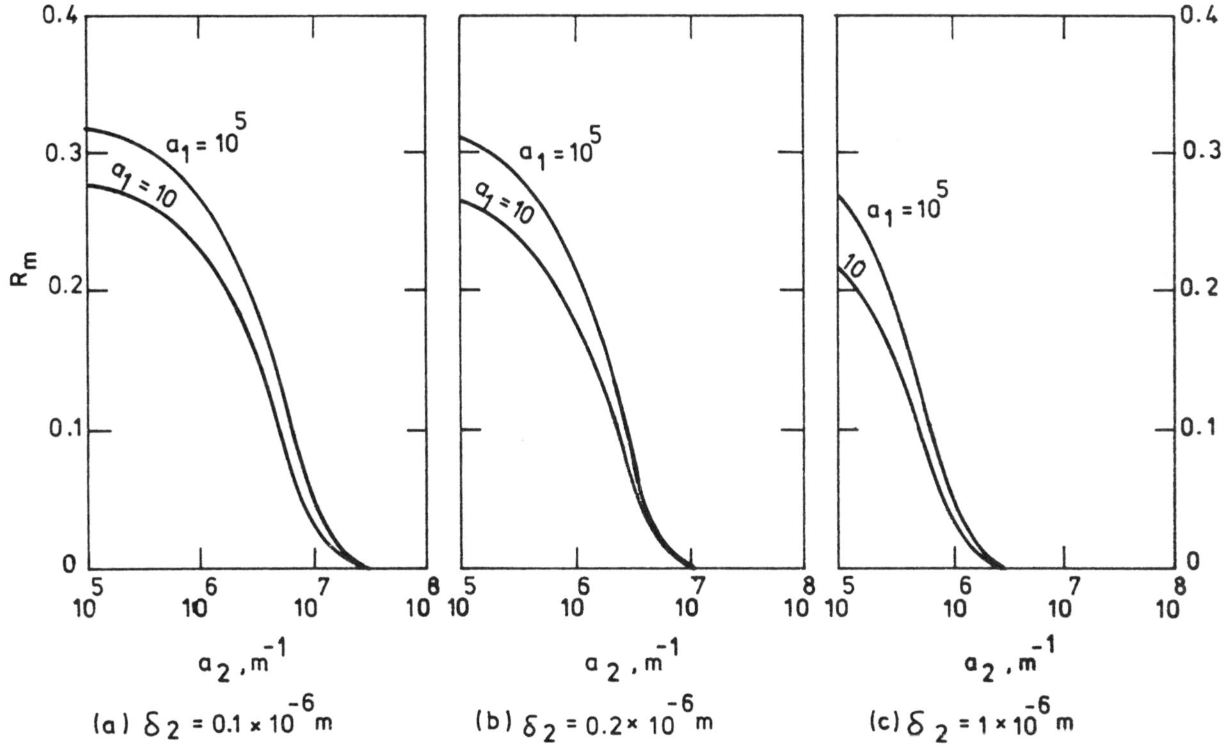

Figure 5 Effect of Absorption Coefficient a_1 and a_2 on Monochromatic Reflectance of the Coating R_m ($\delta_1 = 1 \times 10^{-7}$ m).

LITERATURE CITED

1. Melamed, L. and G.M. Kaplan, "Survey of Selective Absorber coatings for solar Energy Technology," J.Energy 1(2), (March-April 1977).

2. Blickensderfer, R., D.K. Deardoff, and R.L. Lincoln, "Technical Note: Spectral Reflectance of TiN and ZrN Films as Selective Solar Absorbers." Solar Energy, 19, 429 (1977).

THE ORIGIN AND INFLUENCE OF NONEQUILIBRIUM ELECTRIC FIELD EFFECTS DEVELOPED AT PHOTOELECTRODE/ELECTROLYTE INTERFACES

HAROLD E. HAGER

Department of Chemical Engineering
University of Washington
Seattle, Washington 98195

A growing awareness that photo-assisted electrochemical systems may provide the best device technology for some direct solar energy conversion applications is manifested by the rapidly expanding literature in this area. Nevertheless, poor device performance remains a major problem in this research. This is particularly true for many of the photogenerative systems which have been studied. The most widely investigated photogenerative reaction has been the photo-electrolysis of water. Growing evidence suggests that unfavorable photo-assisted water electrolysis cell performances arise, at least in part, from product ion adsorbate effects. The latter appears to contribute both to materials stability and photon utilization efficiency vs. bias problems. This product ion phenomena should also be of importance, in general, for any system incorporating photoelectrodes which utilize depletion space charge fields.

SOLAR COAL GASIFICATION: PLANT DESIGN AND ECONOMICS

A plant design has been layed out in which focused solar energy is used to gasify coal and the resulting synthesis gas is converted into pipeline gas.

The overall performance of the plant and its economics are compared to a similar Lurgi plant. The advantage of the solar coal gasification (SCG) method is conservation of coal, 40% more product can be produced from a given amount of coal. The primary detriment of SCG is the 8 hr/day operation; which leads to a higher plant investment for a given annual production. This factor is twice as important as the cost of the heliostat field. Thus, operating costs are lower but capital costs are higher. In the future, as coal costs increase relative to investment costs; the SCG plant will become more economical than the Lurgi plant.

WILLIAM R. AIMAN
CHARLES B. THORSNESS
and
DAVID W. GREGG
Lawrence Livermore National Laboratory
Livermore, CA 94550

Focused solar thermal energy is a renewable energy source that should be available for future use by the United States. A major R&D effort is underway for producing electricity from this resource. A parallel effort is appropriate for processes that produce fuels. Liquid or gaseous fuels are not only storable but also can be transported over long distances more easily than electricity. In addition, our transportation system requires liquid fuels.

Production of fuels using solar thermal energy can conserve our resources of fossil fuels in a number of ways. Fuel energy can simply be replaced by solar energy to produce process heat. For example, in steam-flooding an oil field about one quarter of the oil is burned to raise the steam. Solar energy can substitute for this oil. In some other processes, use of solar energy can be even more advantageous because of its unique properties. Very high heating rates can be attained and the energy is delivered without adding anything else to the process stream. These properties can result in higher process efficiency.

ACKNOWLEDGEMENT
This work was performed under the auspices of the U.S. Department of Energy, Lawrence Livermore Laboratory under contract No. W-7405-Eng-48.

We are currently engaged in developing two concepts that use focused solar energy to produce fuels from fossil energy resources: Solar Coal Gasification (SCG) and Solar Oil Shale Retorting (SOSR). A preliminary analysis of SCG was made,[1] a field experiment was performed,[2] and a simple model for the gasification process was discussed.[3] Our Solar Oil Shale Retorting investigations have also been reported.[4,5]

These investigations are in preliminary stages. Initial analyses showed that the processes are technically feasible, fossil energy is conserved, and that the economics would probably be favorable. The prediction of favorable economics was based on certain advantages of the proposed solar processes compared to conventional processes.

The purpose of the present paper is to present a possible plant design for SCG and an economic analysis of that design. The plant has a straightforward, simple design: solar energy is focused directly on reacting coal char with no intervening heat transfer stages and the plant operates only when the solar energy is available with no storage schemes considered.

Analysis of a simple configuration is a good starting point because the concept's strengths and weaknesses are revealed.

0065-8812-81-6037-0210-$2.00

Alternative configurations may alleviate weaknesses in SCG without compromising its strengths. Various other concepts are being considered: (1) Indirect gasification using a heat transfer fluid. This allows the gasification reactor to be on the ground separate from the solar reheat furnace. (2) Hybrid gasification using solar energy when available and coal combustion energy at other times. (3) Syn gas storage so that the down-stream processing units can continue to operate through the night. (4) Solar energy storage so that the entire processing plant can continue to operate through the night.

The intended use of this process and economic analysis is to illuminate the differences between solar thermal processing and conventional techniques. The results are sensitive to the assumptions made in the analysis and should be used only for internal comparison purposes. Several of the assumed values could be in error by 50% or more. The tables can be used for recalculations with other assumed values. The process flows are reasonably good estimates. The economic results are not as well founded.

COAL GASIFICATION

Coal gasification involves two basic chemical steps: pyrolysis and char gasification. Coal decomposes when heated in the absence of air:

$$\text{Coal} \longrightarrow \text{Char} + CO + CO_2 + H_2 + CH_4 + \text{tars} \qquad (1)$$

This pyrolysis is essentially complete at temperatures well below char gasification temperatures. The char has various compositions depending on the coal that was pyrolyzed and on the conditions of pyrolysis, for example: $CH_{0.14}N_{0.007}S_{0.004}$ is a pseudo-molecule that represents a char made from Wyodak coal from the Anderson seam in the Powder River Basin. Compared to the char gasification reaction relatively little energy is required for the pyrolysis reaction because the decomposition reactions are only slightly endothermic; the energy required for pyrolysis is primarily that which is needed to heat the coal to the pyrolysis temperature.

Gasification of char occurs by endothermic reactions with H_2O or CO_2 that are similar to the following:

$$C + H_2O \longrightarrow H_2 + CO \qquad \Delta H_{1000K} = 135.85 \text{ kJ/mol} \qquad (2)$$

$$C + 2H_2O \longrightarrow 2H_2 + CO_2 \qquad \Delta H_{1000K} = 100.95 \text{ kJ/mol} \qquad (3)$$

$$C + CO_2 \longrightarrow 2CO \qquad \Delta H_{1000K} = 170.75 \text{ kJ/mol} \qquad (4)$$

These reactions proceed at reasonable rates only above about 1000K. In steam gasification, reaction (2) dominates at high temperatures. Reaction (4) is unimportant if H_2O is present. Reaction (3) is the sum of reaction (2) and the slightly exothermic water-gas shift reaction:

$$CO + H_2O \longrightarrow CO_2 + H_2 \qquad \Delta H_{1000K} = -34.90 \text{ kJ/mol} \qquad (5)$$

Graphitic carbon has been used to represent char in these reactions. Char is not a well defined material and its thermochemistry depends on its conditions of formation. Its chemical activity is higher than that of graphite by a moderate but not well known amount. The higher chemical activity reduces the endothermicity of the char reactions slightly.

Energy must be supplied to drive these gasification reactions. The "wet" coal must be heated, dried, and heated further to attain the gasification temperatures at which the heat of reaction must be supplied. The "bound" water, which is very loosely attached, is removed in the drying process. During the second heating phase, pyrolysis occurs by processes that are about neutral thermally. Char gasification with steam probably occurs via Reaction 2 or Reaction 3. There is more than enough available energy in the hot gasification products from char gasification to heat, dry, and pyrolyze the coal if counter-counter processing is used.[1,12] Thus, the energy requirement amounts to 101 to 136 kJ per mol of char gasified plus an allowance for high temperature heat loss.

In conventinal gasifiers this energy requirement is supplied by burning part of the coal. Air is used to burn the coal in certain applications where all that is required is a combustible gas product. Most applications, however, require a nitrogen-free product. This product is used as a chemical reactant to synthesize methane, methanol, gasoline, ammonia, or a wide range of other products including monomers for plastics. For the gasifier to supply a nitrogen-free product, oxygen must be used rather than air.

Solar coal gasification does not use coal combustion to supply the energy, so oxygen is not required and in addition less coal is needed for a given amount of useful product. These two advantages for solar coal gasification have a major impact on the fundamental economics of the process.

A SIMPLE MODEL OF SCG

A simple model of SCG has been used to match the product composition obtained in a field experiment in SCG.(3) The model is illustrated in Figure 1 and Table 1 lists the product rate equations. The model parameter "x" reflects the relative

Table 1. Product Yield Equations from Solar Coal Gasification Model, mol/mol of coal feed.

Specie		Pyrolysis Products		Char Gasification Products
CO_2	=	0.034	+	x 0.719 F
CO	=	0.031	+	(1 - x) 0.719 F
H_2	=	0.091	+	(0.769 + 0.719 x)F
CH_4	=	0.050	+	0
C_3H_7	=	0.013	+	0
H_2O (Net)	=	0.587	-	0.719 (1 + x) F
H_2O (Gross)	=	0	-	0.719 (1 + x) F
Tar ($CH_{1.4}$)	=	0.128	+	0
Char ($CH_{0.14}$)	=	0.719	-	0.719 F
Ash (g/mol)	=	0	+	(1.1)

importance of Reactions 2 & 3. Parameter "F" is the fraction of the char produced in the pyrolysis front that reacts in the char gasification front. For a moving bed gasifier at steady state, F = 1. The particular constants shown are for subbituminous coal from the Roland seam in the Powder River Basin of Wyoming. Table 2 compares experimental data and the model's results for fixed-bed gasification and also

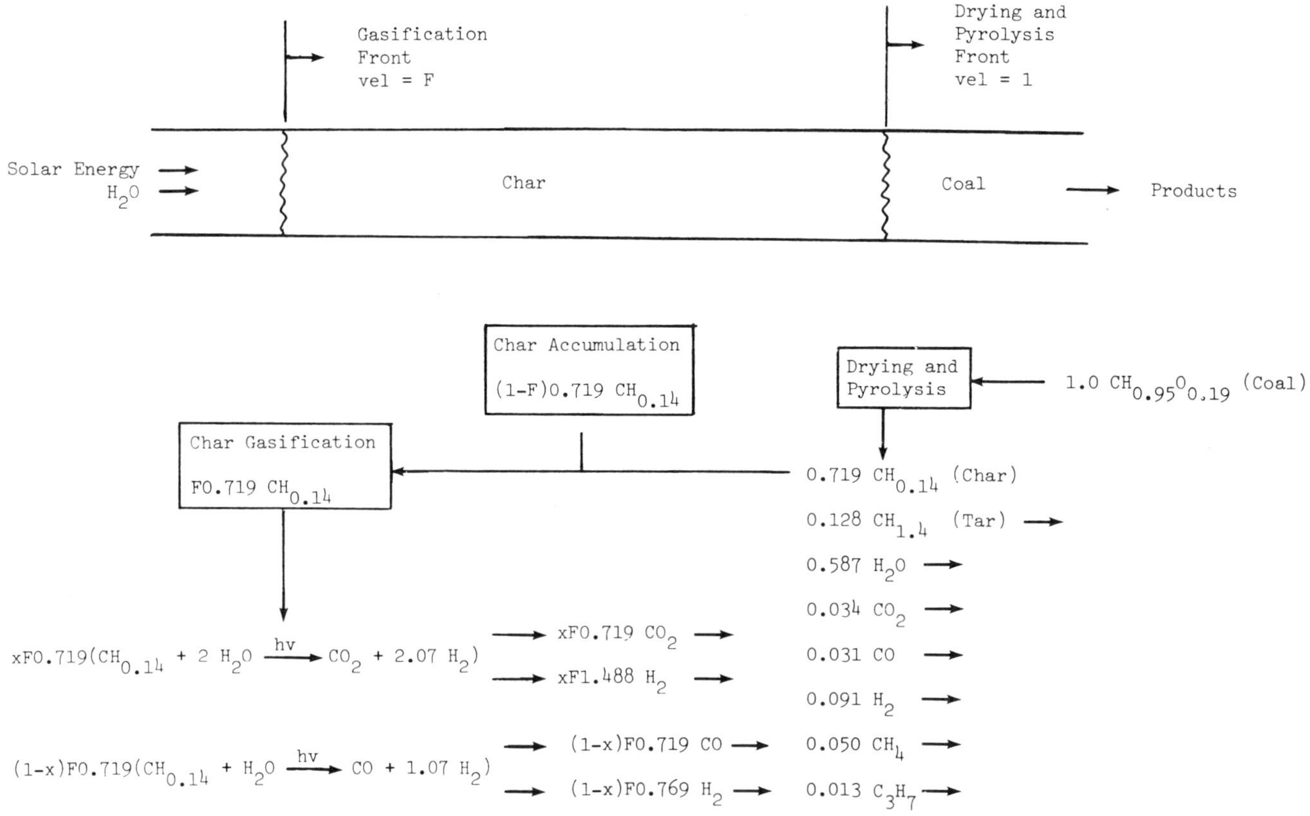

Figure 1. Illustration of Model for Solar Coal Gasification.

Table 2. Solar Coal Gasification Model Results.

Basis: 1 mol of coal feed.	Expt Data	Model Results: Fixed Bed		Model Results: Moving Bed	
Model Parameters:					
F		0.692		1.000	
x		0.518		0.518	
Product Gas:	Mol Frac	mol	Mol Frac	mol	Mol Frac
CO_2	0.205	0.292	0.19	0.406	0.20
CO	0.190	0.271	0.18	0.378	0.18
H_2	0.57	0.881	0.59	1.232	0.59
CH_4	0.0351	0.050	0.033	0.050	0.024
C_3H_7	------	0.013	0.0086	0.013	0.0063
Solids & Liquids:					
Tar ($CH_{1.4}$)		0.128		0.128	
Net Char ($CH_{0.14}$)		0.221		0	
Ash (g/mol)		(0.76)		(1.1)	
Net H_2O		-0.177		-0.504	
Gross H_2O		-0.755		-1.09	
Energetics:*		kJ		kJ	
Char Gasification		59.2		85.5	
Gross Steam		33.2		48.0	
Coal		485		485	
Total of Inputs		577		619	
Product Gas		401		531	
Tar		75.8		75.8	
Net Char		91.5		0	
Total of Products		568		607	
Total Products/Total Inputs		0.98		0.98	
Total Products/Coal		1.17		1.25	
Fraction of Stored Solar Energy in Product Gas		0.15		0.20	

* Based on heats of combustion (HHV) except for the steam and char gasification entries where the energy required to raise steam or to gasify char was used.

Specie	CO_2	CO	H_2	CH_4	C_3H_7	Tar	Char	Ash	H_2O
HHV,kJ/mol	0	283	286	891	2100	592	414	0	(44)

Char Gasification Energy = x 101 + (1-x) 136 kJ/mol

shows the model's results for a moving-bed gasifier. This model has been built into a process analysis code that is used to analyze the plant design.

A PRELIMINARY PLANT DESIGN

A preliminary plant design study has been made for an SCG plant producing nitrogen-free synthesis gas that is converted into methane for substitute natural gas (SNG). This design study was patterned after a Lurgi gasifier SNG plant that was to have been located near Farmington, New Mexico in the Four Corners area. The Lurgi plant design was part of a legal submission that requested a Certificate of Public Convenience and Necessity from the Federal Power Commission, Docket No. CP73-131, filed October 9, 1973.[6] This plant was designed to produce 288.6 MM SCFD of pipeline gas at 954 Btu/scf.

A reasonably complete set of process flow sheets is available in this submission. The design was developed by Stearns-Roger based on technology provided by Lurgi Mineralotechnik GmbH. Lurgi proprietary information was excluded from the submission.

The Stearns-Roger/Lurgi (SRL) design was adapted to incorporate solar coal gasification with several assumptions. The solar coal gasifier performs as predicted by the previously discussed model.[3] All required steam is generated in the gasifier water jackets. The non-solar processing units perform as in the SRL design.

Changes in performance due to transient response are not considered in the analysis. Careful design will be necessary to allow for daily startup and shutdown and there would be energy "losses" in heating up the equipment during start-up. A much more detailed analysis of the plant would be required to allow for these factors, but they should be kept in mind as debits against the SCG plant.

A code was written that calculates process flows. This code is modular in nature and can be applied to many different plant designs. A listing of this self-documented code is available on request. A complete set of flow sheets for this SCG plant is available in Reference 7, the preprint of the present paper. An equivalent analysis was made of the SRL plant for comparison.

PLANT LAYOUT

A schematic of the solar coal gasification/SNG plant is shown in Figure 2. The functions of all of the areas of Figure 2 are quite conventional except for areas 33 and 35 which are the Heliostat Field and the Gas Generator.

Coal is crushed, screened, cleaned, and stored in Area 25. This coal is fed into the Gas Generator, exposed to the concentrated solar flux, and gasified in

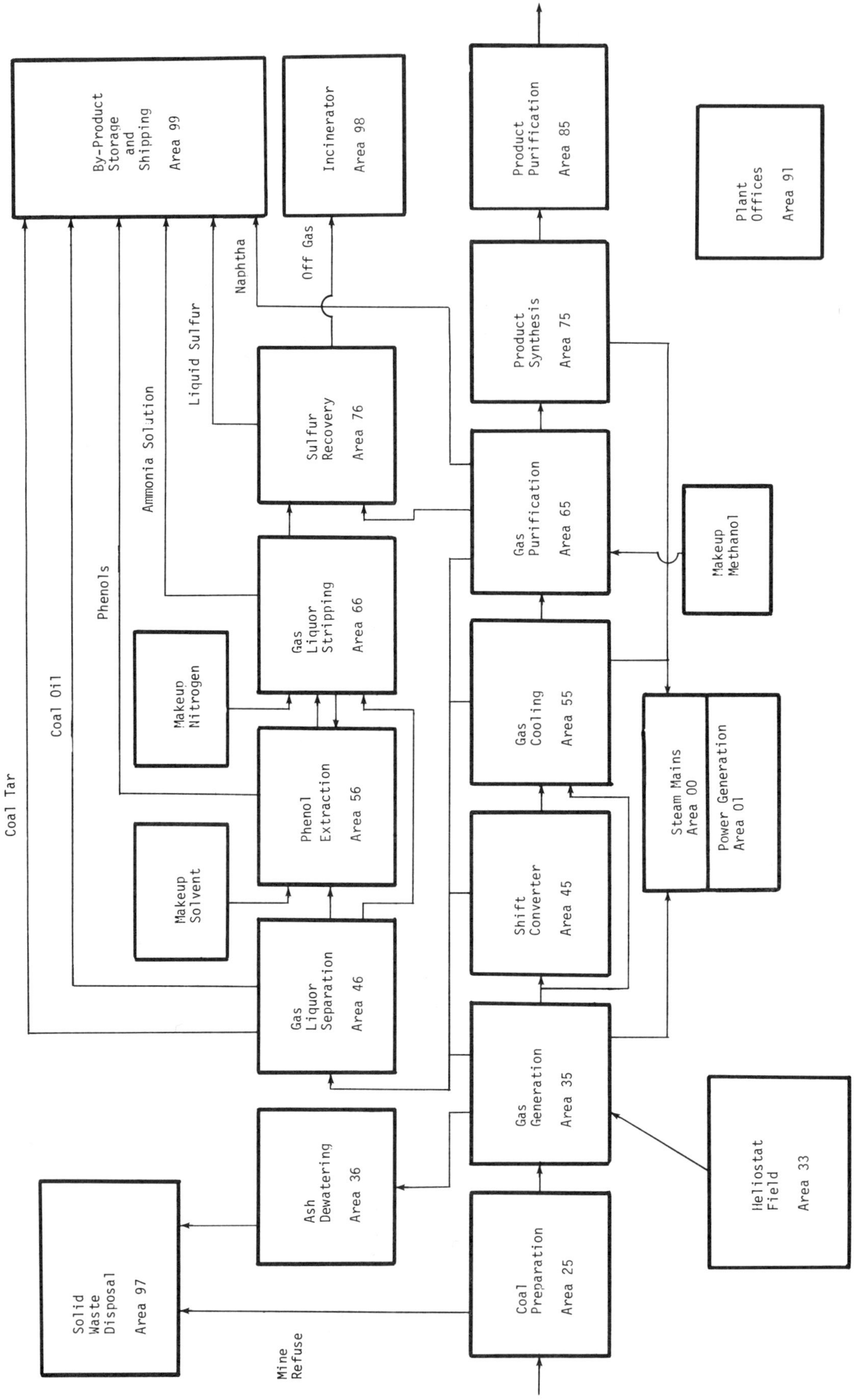

Figure 2. Solar Coal Gasification Plant Layout.

Area 35 which is adjacent to Area 33, the Heliostat Field. Part of the crude gas is fed into the shift converter to increase the hydrogen content. The other portion of the crude gas bypasses the shift converter. The bypass ratio is adjusted so that when the two streams are reblended, the composition will match the requirements of the methane synthesis reactor. Because of the high H_2 content of the SCG syn gas, 88% could be bypassed around the shift converter compared to 55% for the Lurgi.

Liquid organic compounds and NH_3 are removed from the gas in Areas 35, 45, 55, and 65. In addition to removing the last of the liquid organic compounds, Area 65 removes the acid gases, H_2S and CO_2, with a chilled methanol wash. This is the Lurgi Rectisol Process. The cleaned and shifted feed gas enters the methane synthesis reactor. The product of this reactor is compressed, dehydrated, and injected into the Product Pipeline.

By-products are separated from the main stream and processed in peripheral areas. These products have substantial value depending on the marketing circumstances, but in any case the plant must dispose of them. The by-products consist of coal tar, coal oil, naptha, phenols, ammonia, liquid sulfur, carbon dioxide, and the fines separated in the coal preparation areas.

Detailed flow sheets are available for all of these processing units in the preprint of the present paper.[7] The performance of each unit, other than the gasifier, is assumed to be similar to those in the SRL design with no adjustments for size This assumption is reasonable because these plants are composed of multiple units and size variations are accomplished by adding or subtracting units rather than by changing sizes. For example, in the SRL design there are 28 Lurgi gasifier units.

PLANT PERFORMANCE SUMMARY

The overall performance of the solar coal gasification plant can be obtained by comparing the input and product streams as in Table 3. The flow of energy is shown for various important streams. This energy flow is the "cold gas" heat of combustion except for the solar energy and steam streams.

Table 3. Coal Gasification Plant Performance.

	Solar	Lurgi
(1) Coal Energy (excluding coal fines), MJ/s	12370	5271
(2) Steam (Heat of Vaporization), MJ/s	1335	548.7
(3) Solar Energy, MJ/s	6154	NA
(4) Clean Syn Gas Heating Value, MJ/s (excluding NH_3 & H_2S)	13557	3853
(5) Organic Liquids Heating Value, MJ/s	1728	309.4
(6) Ammonia Heating Value, MJ/s	127.4	49.95
(7) Hydrogen Sulfide Heating Value, MJ/s	55.68	28.18
(8) Char Heating Value, MJ/s	nil	nil
(9) Total Gasifier Product Heating Value, MJ/s	15470	4240
(10) Stored Solar Energy, MJ/s (10 = 9 - 1)	3094	NA
(11) Product Gas Storing Fraction, % (11 = (10/9)x100)	20	NA
(12) Solar Energy Storing Efficiency, % (12 = (10/3)x100)	50	NA
(13) Gasifier Thermal Efficiency, % (13 = (9/(1 + 2 + 3))x100)	80	80
(14) Finished Product HHV, kJ/mol	851.8	855.7
(15) Finished Product Rate, kg/s	209.9	63.07
(16) " " " kgmol/s	12.89	3.921
(17) " " " MJ/s	10980	3355
(18) " " " 10^{15} J/yr*	104.3	104.3
(19) " " " 10^{12} Btu/yr	98.86	98.86
(20) Finished Product Energy/Coal Energy, %	88.76	63.65

* SCG - 8 hr/day, 330 day/yr
Lurgi-24 hr/day, 360 day/yr

Sized coal with a heat combustion of 12370 MJ/s is gasified with steam (heat of vaporization energy of 1335 MJ/s) by use of solar energy (at 6154 MJ/s). Heat loss from the gasifier of 20% of the total input is assumed. This heat loss matches the heat loss from the Lurgi gasifier. The gasifier product consists of clean syn gas -- 13557 MJ/s, organic liquids -- 1728 MJ/s, ammonia -- 127.4 MJ/s, and hydrogen sulfide -- 55.68 MJ/s. The total gasifier product heat of combustion is 15470 MJ/s.

The amount of stored solar energy (3094 MJ/s) is the total product energy (15470 MJ/s) minus the coal energy (12370 MJ/s). The steam requirements can be met by raising steam in water jackets around the gasification reactors using part of the 20% heat loss (3972 MJ/s).

The Product Gas Storing Fraction (fraction of the product energy which is stored solar energy) is 20%. (See Table 3.) Solar energy is treated here as if all of the heat loss is supplied by the solar energy and none by the coal energy. The energy efficiency (total product energy divided by total input energy) is 80%, which is a direct consequence of the 20% heat loss assumed in order to match the

Lurgi units. The Solar Energy Storing Efficiency is the Stored Solar Energy divided by the Total Solar Energy. In this case it is 50%.

The ability to convert 50% of the incoming solar energy into chemical energy in the syn gas is remarkable. This efficiency compares to the efficiency of an electric plant which is below about 40%.

AN ECONOMIC ANALYSIS

An economic analysis has been made of the solar coal gasification plant described in the previous section. This analysis was based on the cost data released in the same legal submission(6) that provided the data for the flow sheets. As in the calculations for the flow sheets, it is assumed that modest changes in the size of the plant will have little effect on the costs per unit of production rate. The cost data was updated from 1973 to 1978 to compare to 1978 heliostat cost data using the Nelson Refinery Cost Index and Chemical Index.(7) The cost ratios were 1.498 and 1.635 respectively.

CAPITAL COSTS

For each of the major processing areas (Figure 2) the cost was assumed to be proportional to the production rate of the area's product (Table 4a). In sizing the units to give the same annual production rate, 8 hr/day and 330 day/yr was used for SCG and 24 hr/day and 360 day/yr was used for Lurgi. The shorter year allows for bad weather.

The cost for the gasifier area itself can only be roughly estimated at this time since only general ideas of its design are known. The simplest of the current ideas(1) include a moving bed with coal surface exposed to solar energy at the top. This moving bed arrangement is somewhat like that of the Lurgi gasifier, so the costs may be similar. The Lurgi design requires drive shafts that penetrate the pressure shell into the high temperature zones. These drive shafts would not be required in the SCG gasifier, which would reduce the cost. However, a window to admit the solar energy will be required, which would increase the cost. For the purpose of the present analysis, the cost of the gasifier area per unit production rate will be assumed to equal that for the SRL plant.

The effect of errors in estimating the gasifier cost can be estimated as a fraction of the other costs. The gasifier cost is less than 20% of the total cost for the plant including the heliostat field, and the capital costs are less than 60% of the cost of the final product ($2/gJ coal cost); so less than 12% of the product cost is from the gasifier capital costs. A 50% error in estimating the gasifier cost becomes less than a 6% error in the product cost.

The cost of the heliostat field to supply the solar energy was estimated from information developed for the DOE Prototype Heliostat program. The energy collected and costs for a McDonnel Douglas heliostat were used.(8) These 49m^2 heliostats will collect 368 gJ/yr under Barstow conditions and will cost $3136 (1978$) apiece installed (Table 4b).

The total capital required for the processing plant is 1439 M$ for SCG and 465 M$ for Lurgi (Table 4c). The solar facility for SCG would be an additional 553 M$. The additional processing plant investment for SCG is due to oversized equipment that compensates for the 8 hr/day operation. The additional processing plant investment is approximately twice the cost of the heliostat field.

OPERATING COSTS

The annual operating costs consist of Raw Material Costs, Labor and Maintenance, Local Taxes and Insurance, Depreciation, and annualized costs for the capital investment minus the by-product credits. The Annual Operating Costs and Product Costs are shown in Table 4d. Allowing for by-product credits is difficult because marketing conditions vary with plant location. For example, in some locations the CO_2 will also have value for flooding oil fields in tertiary recovery operations or for other purposes. The values of the by-products were obtained from Chemical Marketing Reporter.(9)

For the coal cost of $2/gJ used in Table 4d, Total Operating Cost is 429 M$/yr for the SCG plant and 368 M$/yr for the SRL plant.

PRODUCT COSTS

These plants produce 104.3 10^{15} J/yr of product energy, so the product cost is

Table 4a. Processing Plant Investment.

	units	Solar	Lurgi
COAL HANDLING			
Crushing, Screening, Blending, and Storage	k$	47787	21977
Lock Gas Storage and Compression	k$	4033	1855
Total Direct Cost--Coal Handling	k$	51820	23832
PROCESSING UNITS			
Gas Production	k$	293921	97121
Shift Conversion	k$	8929	11064
Gas Cooling	k$	43636	11841
Gas Purification	k$	201493	45839
Refrigeration	k$	62480	14214
Product Synthesis	k$	106619	27228
Product Purification & Final Preparation	k$	50382	12867
Gas Liquor Separation	k$	11643	2153
Phenol Extraction	k$	45113	8468
Ammonia Stripping	k$	19115	7496
Sulfur Recovery	k$	19744	9993
Total Direct Cost of the Processing Units	k$	863075	248283
UTILITY PLANTS			
Fuel Gas Production & Treating	k$	0	40316
Air Compression	k$	0	30668
Oxygen Production & Compression	k$	0	43343
Steam & Power Generation	k$	25402	44591
Cooling Water System	k$	5289	8699
Raw Water System	k$	6130	10078
Miscellaneous Plant Utility Systems	k$	1842	7403
Total Direct Cost--Utility Systems	k$	38662	185148
ASH HANDLING AND RAW WATER SYSTEMS			
Ash Dewatering & Transfer	k$	4443	2043
Water Pumps, Pipeline, & Storage Ponds	k$	12960	21312
Total Direct Cost-- Ash Handling & Raw Water Systems	k$	17403	23355
Total Direct Cost of the Processing Plant Including Contractor Engrg.& Fees, Licenses & State Tax with no Contingencies	k$	931140	302873

Table 4b. Solar Facility Investment.

Solar Energy Collection Rate	kW/Heliostat	35
Cost of Heliostats (installed & wired)	k$	3.136
Land Required	sq meters/heliostat	245
Solar Energy Required	MW	6154
Number of Heliostats Required		175835
Total Heliostat Cost	k$	551419
Land for Heliostat Field	sq mi	16.63
Price for Land	k$/sq mi	100
Cost of Land for Heliostat Field	k$	1663
Total Cost of the Solar Facility	k$	553083

Table 4c. Summary of Investment Requirements.

	units	Solar	Lurgi
1. Total Direct Cost of the Processing Plant including contractor engg & fees, licenses & state tax with no contingencies	k$	934140	302873
2. Initial Catalysts and Chemicals	k$	38948	9946
3. Cost of Land for Processing Plant	k$	457	151
(Land for processing plant)	sq mi	4.57	1.51
4. Site Improvements & General Facilities	k$	157204	51945
5. Funds for Use During Construction (20% of the cost of the processing plant)	k$	186828	60575
6. Startup Costs (8% of the cost of the processing plant)	k$	74731	24230
7. Working Capital	k$	46707	15144
8. Total Capital Required for Processing Plant (8 = 1 + 2 + 3 + 4 + 5 + 6 + 7)	k$	1439014	464864
9. Total Cost of the Solar Facility	k$	553083	none
9a Total heliostat cost	k$	551419	none
9b Cost of land for heiostat field	k$	1663	none
10. Total Capital Investment (10 = 8 + 9)	k$	1992097	464864
11. Total Depreciating Investment (Subtracting: land costs, working capital, initial catalysts, and chemicals) (11 = 10 - (3 + 9b) - 7 - 2)	k$	1904322	439623
12. Total Plant Investment (Subtracting: working capital, startup costs, initial catalysts and chemicals) (12 = 10 - 2 - 6 - 7)	k$	1831711	415544

$4.12/gJ for the SCG plant and $3.26/gJ for the SRL plant with coal at $2/gJ.

The cost of the coal affects the product cost significantly (Figure 3). The Lurgi plant has lower product cost up to a coal cost of about $4.00/gJ. Above this crossover point the lower coal requirement for the solar gasifier results in lower costs.

The advantages and disadvantages of SCG are well summarized in Figure 3. The initial disadvantage of higher investment requirement is reflected by the intercept on the vertical axis. The way in which this initial disadvantage can be overcome at high coal costs is shown by the way the cost lines cross. The different slopes reflect the different coal requirement per gJ of product energy.

Table 4d. Annual Operating Cost & Product Cost.

RAW MATERIALS	units	Solar	Lurgi
Raw Coal Cost	k$/yr	255688	356390
Cost of Raw Coal	$/gJ	2.00	2.00
Catalysts and Chemicals	k$/yr	11408	9532
Water Costs	k$/yr	938	1014
1. Total Cost of Raw Materials	k$/yr	268034	366937
LABOR COSTS			
A. Operating Labor	k$/yr	7812	7751
Operating Labor Force		313	310
Labor Pay Rate	k$/yr	25	25
B. Supervisory Labor--20% of (A)	k$/yr	1563	1550
C. Overhead--60% of (A)	k$/yr	5629	5580
D. Supplies--30% of (A)	k$/yr	2345	2325
2. Total Labor	k$/yr	17355	17206
3. Maintenance (3% of the Direct Cost of the Processing Plant Plus 1.5% of the Total Cost of the Solar Facility)	k$/yr	36320	9086
4. Local Taxes and Insurance (1.5% of the Total Plant Investment)	k$/yr	27476	6233
5. Gross Operating Cost (5 = 1 + 2 + 3 + 4)	k$/yr	349185	399463
BY-PRODUCT CREDITS			
A. Liquid Sulfur	k$/yr	1507	2496
B. Ammonia	k$/yr	7001	8985
C. Coal Tar	k$/yr	9485	5446
D. Coal Oil	k$/yr	59015	36220
E. Phenols	k$/yr	47165	28970
F. Naphtha	k$/yr	39968	21864
G. Coal Fines	k$/yr	10228	15392
6. Total By-Product Credit	k$/yr	174367	119372
7. Net Operating Cost (7 = 5 - 6)	k$/yr	174818	280091
8. Depreciation (5%/yr of Total Depreciating Investment)	k$/yr	95216	21981
9. Allowance for: Interest on Debt, Return on Investment, and Income Tax (8% of Total Capital Investment)	k$/yr	159368	37189
10. Total Operating Cost (10 = 7 + 8 + 9)	k$/yr	429402	339261
11. Total Product Energy 10^{15}	J/yr	104.3	104.3
--Total Flow Rate of Product Gas	kgmol/s	12.89	3.92
--HHV of Product Gas	kJ/gmol	852	856
12. Product Cost (12 = 10/11)	$/gJ	4.12	3.26

Some uncertain factors in the gas cost are the annuallized cost of capital and allowance for by-product value. Annuallized capital cost covers interest on debt, income tax, and return on investment. The rate used for this cost was 8%. This rate is appropriate for an economy with mild inflation. The effects of higher rates with and without by-product credits are shown in Table 5.

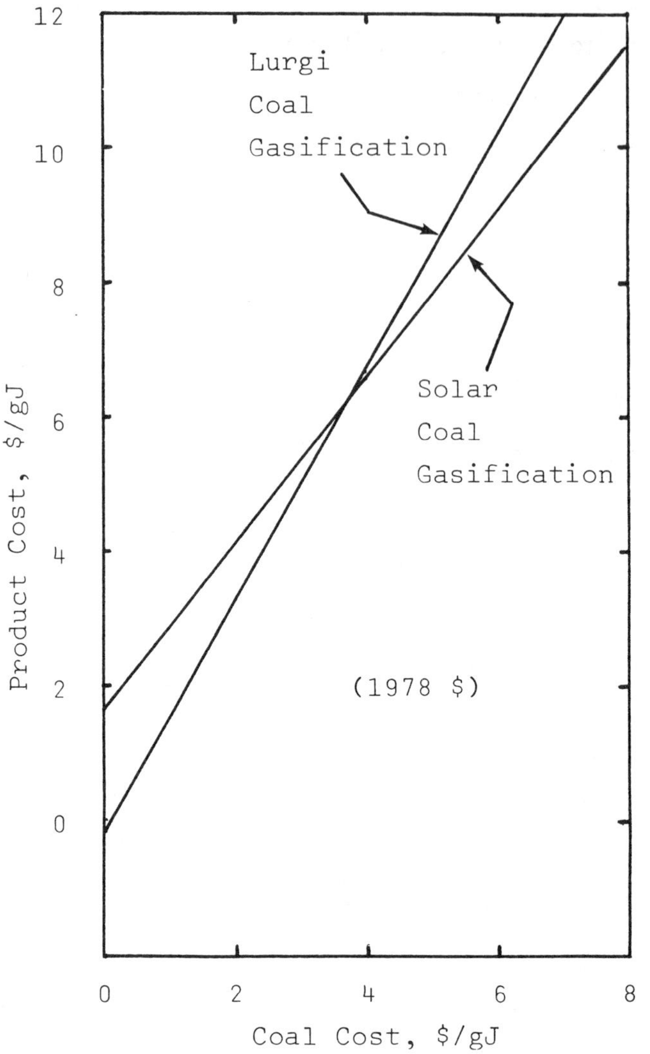

Figure 3. Product Cost for Solar Coal Gasification and for Lurgi Coal Gasification as a Function of Coal Cost.

Other estimates[10] have shown higher product costs than the estimates shown in Figure 3 for both solar and Lurgi coal gasification. The present results were intended to be used for internal comparison only. Resolution of differences between the sources of data is beyond the scope of the present paper. The internal consistency in the present results should be adequate for comparison of solar coal gasification with Lurgi coal gasification. This paper is not intended to be the final answer on SCG costs. The assumed values for a number of the parameters could be different from those the reader may prefer. The installed cost of the heliostats is an open question. The value of the byproducts

Table 5. Gas Costs with and without By-Product Credits and with the Annuallized Capital Cost (ROC) at 8, 16, & 24% of the Total Capital Investment. (1978$)

Coal Cost $/gJ	With By-Product Credits: Gas Cost SCG $/gJ	With By-Product Credits: Gas Cost Lurgi $/gJ	Without By-Product Credits: Gas Cost SCG $/gJ	Without By-Product Credits: Gas Cost Lurgi $/gJ
With ROC = 8%				
0	1.66	-.15	3.33	0.98
1	2.89	1.55	4.56	2.68
2	4.12	3.26	5.79	4.39
4	6.57	6.68	8.24	7.81
With ROC = 16%				
0	3.18	0.21	4.85	1.34
1	4.41	1.91	6.08	3.04
2	5.64	3.62	7.31	4.75
4	8.09	7.04	9.76	8.17
With ROC = 24%				
0	4.70	0.57	6.37	1.70
1	5.93	2.27	7.60	3.40
2	7.16	3.98	8.83	5.11
4	9.61	7.40	11.28	8.53

depends on marketing factors. The required rate of return on the investment is extremely uncertain in these times. The format of the results allows recalculation using other assumed values for these parameters.

The only hard information on costs comes from actually building a plant and operating it for a period of many years. All efforts at economic analysis give results that are soft to some degree. The results of our analysis are based on an adaptation of a particular plant design to SCG. A rather modest level of effort was available to make this analysis. As development of the SCG concept proceeds, better plant designs and better economic analysis will be required.

Even the straightforward, simple plant layout analyzed here is not unreasonably expensive compared to a Lurgi plant; in addition, minor changes in the process performance or cost parameters can have a major effect on the location of the crossover point. The location of this point is sensitive to small changes and modest improvements in the SCG process may move the crossover down into the range of current coal costs.

The largest detriment for SCG is the 8 hr/day operation. This detriment is the main cause of the high capital requirement for a given annual production. Alleviating this detriment should be the first goal of further work on SCG.

SUMMARY

The present paper is one further step in the process of developing the concept of Solar Coal Gasification. A number of problems were revealed and some advantages were quantified.

A plant was layed out and analyzed with the aid of a code that calculates process flows and plant economics. This plant is the simplest, most straightforward plant and thus it was the most appropriate for the initial analysis. Solar energy was focused directly on the reacting coal and no energy storage schemes were considered in this analysis. Various improvements on this simple scheme can be made.

The process analysis showed that 40% more product can be produced from a given amount of coal and that the fraction of the solar energy that can be converted into usable energy is greater than the fraction for any other method with the exception of conversion to simple thermal energy.

The economic analysis showed that for a plant that produces a given amount of product per year, the capital costs are much higher but the operating costs are much lower. These cost differences reflect the 8 hr per day operation for the SCG plant plus the cost of the heliostat field, and the lower coal consumption for the SCG plant.

Product costs were derived as a function of coal cost. The Lurgi plant was more economical when coal was less than about $4/gJ. The SCG plant was more economical when coal was more than $4/gJ. Currently Coal Week shows most coal costs between $1/gJ and $2/gJ. Mine-mouth coal is cheaper than the coal listed in Coal Week.

Eventually coal prices may rise to a level (relative to plant costs) where the simple, straight-forward SCG plant will be cheaper than a Lurgi plant. In the mean-time we should look for better SCG concepts. Alternate concepts being considered for economic and practical reasons include:

(1) Indirect gasification using a heat transfer fluid.
(2) Hybrid gasification using solar energy when available and coal combustion energy at other times.
(3) Syn gas storage so that the down stream processing units can continue through the night.
(4) Solar energy storage so that gasification can continue through the night.

In the early stages of developing a new concept, each step is very likely to reveal more problems than the step has cleared up. We are certainly in this position now. Analysis of the straightforward plant design has revealed many problems. We will have to be more clever to get a design that can be considered practical.

CONCLUSIONS

It seems clear that SCG will become competitive economically if the price of coal increases sufficiently compared to plant costs, because SCG conserves coal. In addition, since it is in the national interest to conserve coal; use of SCG might be subsidized.

LITERATURE CITED

1. D.W. Gregg, W.R. Aiman, H.H. Otsuki, and C.B. Thorsness, "Solar Coal Gasification," Solar Energy, Vol, 24, pp 313-321, 1980. Also Lawrence Livermore National Laboratory Report UCRL-81853 Rev. 1, Oct., 5, 1979.

2. D.W. Gregg, R.W. Taylor, J.H. Campbell, J.R. Taylor, and A. Cotton, "Solar Gasification of Coal, Activated Carbon, Coke, and Coal Biomass Mixtures," Solar Energy, Vol. 25, pp 353-364, 1980. Also Lawrence Livermore National Laboratory Report UCRL 83440, Sept. 26, 1979.

3. W.R. Aiman and D.W. Gregg, "Gasification of Coal with Solar Energy," 88th National Meeting of the American Institute of Chemical Engineering, Philadelphia, June, 1980. Also Lawrence Livermore National Laboratory Report UCRL-84458, June 1980.

4. D.W. Gregg, J.Z. Grens, R.W. Taylor, and W.R. Aiman, "Solar Retorting of Oil Shale", 88th National Meeting of the American Institute of Chemical Engineering, Philadephia, June 1980. Also Lawrence Livermore National Laboratory Report UCRL-84142, June, 1980.

5. D.W. Gregg, J.Z. Grens, R.W. Taylor and W.R. Aiman, "Solar Retorting of Oil Shale," Proceedings of the 15th Intersociety Energy Conversion Engineering Conference, Seattle WA., August, 1980.

6. Application of El Paso Natural Gas Company for a Certificate of Public Convenience and Necessity, before the Federal Power Commission, Docket No. CP73-131, Second Supplement, filed: October 9, 1973.

7. W.R. Aiman, C.B. Thorsness, and D.W. Gregg, "Solar Coal Gasification: Plant Design and Economics", 73rd Annual American Institute of Chemical Engineers Meeting, Chicago, November, 1980. Also Lawrence Livermore National Laboratory Rept. UCRL-84610, November, 1980.

8. "Nelson Cost Indexes", Oil and Gas Journal June 2, 1980.

9. "Recommendations for the Conceptual Design of the Barstow, CA., Solar Central Receiver Pilot Plant -- Executive Summary," Sandia Laboratories, Livermore, CA., Report No. SAND 77-8035 October, 1977.

10. Chemical Marketing Reporter, March 31, 1980.

11. S. Lakshmann, F. K. Manasse, and V. K. Mathur, "Production of Fuels from High Temperature Solar Thermal Systems-Economic Analysis", Fundamentals and Applications of Solar Energy: AIChE Symposium Series 198, Vol. 76, New York, 1980.

12. C.B. Thorsenss and R.B. Rozsa, "In-Situ Coal Gasification: Model Calculations and Laboratory Experiments," J. Soc. Pet. Eng., April, 1978, 105-116.

A CONTINUOUS TWO STAGE SOLAR COAL GASIFICATION SYSTEM

The use of solar thermal energy has the potential to play an important role in the production of fuels and chemicals. A continuous two stage fluidized bed hybrid solar coal gasification system designed at the University of New Hampshire to produce synthesis gas from various feed stocks—coal, lignite and peat is discussed. Heat and material balance are presented. The experimental data obtained from this bench scale unit will be useful for the commercialization of this technology.

Economics of commercial solar gasification plants with reference to their size are discussed. A suitable capacity for a solar hybrid plant is estimated to about 1.42 $\times$ 10^6 m^3 (50 MM SCF) per day of SNG.

V. K. MATHUR
S. LAKSHMANAN
F. K. MANASSE
V. VENKATARAMANAN
and
R. W. BREAULT

The availability of high temperature solar facilities which have unique capabilities for providing high temperatures, up to 4000°K, and high heat fluxes has led to new concepts of utilizing solar energy and its storage. Many investigators have demonstrated the application of high temperature solar heat for the dissociation of SO_3, oxidation of atmospheric nitrogen, pyrolysis and gasification of coal and biomass. These processes not only produce fuels and chemicals but can also be considered as techniques for storing solar energy in a chemical form.

The use of solar thermal energy in the production of fuels has the potential to reap major benefits with respect to the nation's critical energy situation. Our dependence on foreign oil imports can be reduced, the impact of declining domestic oil and gas production can be lessened, the attractiveness of coal as a feedstock for the production of liquid and gaseous fuels can be enhanced, and the adverse environmental effects of increasing coal usage can be mitigated. Additionally, the conversion of solar thermal energy to fuels represents an ideal solution to one of the major prerequisites for widespread use of solar energy - a means for convenient, long-term, and readily transportable solar energy products.

R.W. Breault is presently with the Department of Chemical Engineering, University of New Hampshire, Durham, New Hampshire 03824.

Literature Review

Gregg et al. (1,2) have conducted experiments at the White Sands solar facility on the gasification of bituminous coal using a directly heated gasifier. They have reported the production of gaseous product with an energy content twenty percent higher than in the initial coal feed. The increase represents the stored solar energy.

In our earlier papers (3,4) we have proposed a gas recirculation hybrid system for solar gasification. A similar system was also proposed by Copeland (5) of the Solar Research Institute, Denver, Colorado. This system consists of a heat source and a gasifier where synthesis gas ($CO + H_2$) is produced, a part of which is methanated to produce pipe line gas (CH_4). The other part of the gas which acts as a heat transfer medium is recycled, preheated either by a heat source fired by solid fuels or by a solar source during the day. The hot gases are fed to the gasifier to supply the necessary heat for the endothermic reaction of steam with the carbonaceous material to form CO and H_2. An economic analysis of this system has shown that by 1995 the cost of SNG produced from a solar hybrid gasification plant will be competitive with the gas produced from conventional gasification units. Antal (6) of Princeton University has also been conducting experiments in a bench scale solar reactor for the pyrolysis of biomass.

0065-8812-81-4865-0210-$2.00

UNH Two Stage Fluidized - Bed Hybrid Coal Gasification System

At the University of New Hampshire a two stage fluidized bed hybrid coal gasification system has been designed to produce synthesis gas (or "Sun-Fuel") from various feed stocks - coal, lignite and peat. With reference to Fig. 1 coal or lignite is devolatilized in the Stage I reactor. The heat necessary to bring the coal to the devolatilization temperature of 823°K is supplied by the recirculating gas heated by the solar receiver/coal heater. The volatile products and the char obtained from the Stage I reactor are fed along with steam and recirculating hot gas to the Stage II reactor where the gasification of carbon with steam takes place at 1227°K. The necessary heat for the endothermic reaction

$$C + H_2O \rightarrow CO + H_2 \; ; \; \Delta H = 134.3 \times 10^3 \text{ kJ/kg mole}$$

is supplied by the recirculating gas entering the reactor at 1339°K.

In most conventional coal gasifiers the product gas contains tars and other condensable products which require cleaning of the gas before its utilization. In the two stage hybrid gasification system that we have proposed the devolatilization of coal (removal of volatiles and tar) is carried out in Stage I. Their subsequent exposure to the high bed temperature in Stage II causes the elimination of both the tar and light oil by cracking (7,8). The resultant product gas is thus free of tar and other condensables. Consequently, the clean up of the product gas for tarry material, which contributes considerably to the capital investment and operating cost, is not required. The sulfur in coal which will be present as H_2S in the product gas due to hydrogenation can be removed by hot clean up processes. Thus the need for quenching the product gas is eliminated resulting in better thermal efficiency of the system.

Since coal is not burned with oxygen to provide the heat for the endothermic reaction(s) the CO_2 content of the product gas is much less than in the gas produced from Lurgi or other conventional gasifiers. Further, since the product gas contains about 50% H_2

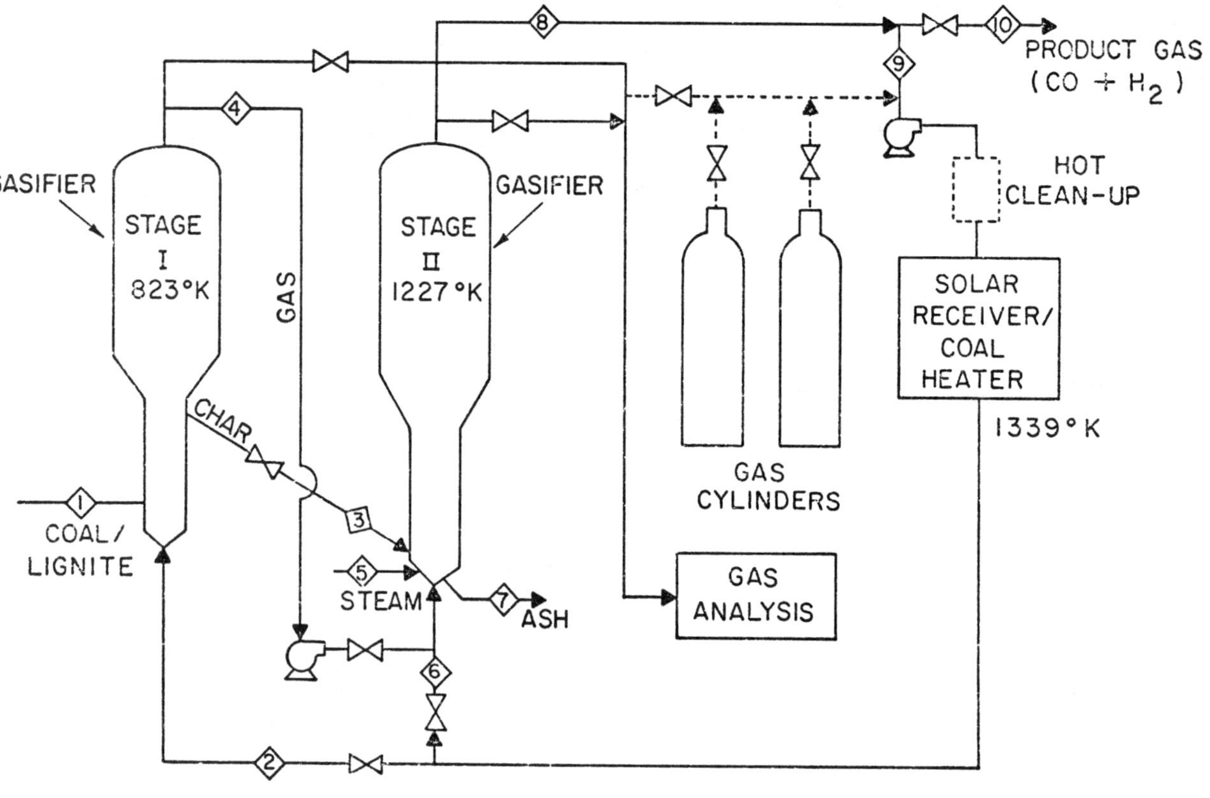

Fig. 1 Schematic Diagram of Bench-Scale Solar-Hybrid Coal Gasification System

(2) the amount of CO_2 produced in the subsequent "shift" conversion to produce the required hydrogen to carbon monoxide ratio for the methanation reaction is very much decreased. This results in a lower capital investment and operating cost for the acid gas purification system. This reduction in carbon dioxide production is of great significance from an atmospheric pollution point of view and makes the usage of coal even more attractive.

Experimental Bench Scale Unit

A bench scale unit for the solar gasification of coal or lignite based upon the concept discussed earlier is under fabrication at the University of New Hampshire. The fluidized bed reactors are designed to operate at 1034 kPa gauge (see Figure 1). Stage I reactor operates at 823°K and consists of a 0.1 m I.D. chamber surrounded by two layers of refractory - fire brick and insulating refractory. The feed to Stage I reactor is dried coal of 200 microns size. The hot gases which also serve as the fluidizing medium enter at a linear velocity of about 0.3 m/sec. The upper portion of the reactor is 0.2 m in diameter and acts as a disengaging zone where the gas velocity is reduced and entrained fines fall back into the reactor bed.

The Stage II reactor is similar in design to Stage I. The I.D. of this reactor is 0.36 m with a disengaging zone of 0.71 m diameter. The temperature in the Stage II reactor is 1227°K.

Industrial size blowers suitable for operation at temperatures up to 1422°K are available in the market. Surprisingly, for this type of operation small size blowers for bench scale units are difficult to obtain. The Stage I reactor acts as a carbonizer whereas the Stage II reactor gasifies the char from Stage I.

The outlet gas stream <8> (see Fig. 1) from Stage II is divided into product stream <10> and a recirculation stream <9> for stages I and II, to be heated to 1339°K in a solar receiver during the day and coal heater at night. The recirculation gases can also be heated by the coal heater during cloudy days thereby preventing any interruption in the product gas supply. This aspect of our scheme, which permits the continuous production of gas, will be of major importance for the commercialization of this technology because it overcomes the major drawback of solar energy namely, its intermittent availability.

Heat and Material Balances

A review of the literature reveals that there is no research work on the gasification of coal using hot synthesis gas. The data available on the Tri-Gas Gasification Process (8) which is a conventional gasification system using three stages is limited. Even if more complete data were available on the system, it would be of very little value since the two systems, namely Tri-Gas and the proposed hybrid scheme are very different. However, for properly sizing the reactors, pipe lines, heaters and blowers for our bench scale unit, it was essential that we make heat and material balances. Estimated values of such balances are presented in Table I through IV and are based on 80% coal conversion and gas compositions as obtained with conventional coal gasification systems (9). The unconverted coal from the gasifier is used as a heat source in the coal heater. Consequently no carbon is lost in the process. How far these balances will differ from the actual measured values will be a major purpose of our experimental work.

The heat and material balances as presented do not indicate any conceptual or practical problem that we may encounter in performing the experimental work or scaling the unit to larger size if so desired.

Objectives of the Bench Scale Experiments

The major objectives of our experimental work at the University of New Hampshire are as follows:

(a) Design, fabrication and operation of a gas recirculation hybrid solar gasification unit.

(b) Heat and material balances.

(c) Study of heat transfer rates for the carbonizer and gasifier.

(d) Study of gas conversion rates for the carbonizer and gasifier.

(e) Optimization of design parameters viz. temperatures of the recirculation gas and reactors, pressure, space velocity for the gases and composition.

In addition to the objectives mentioned above, one of the most important purposes of

Table I. Stage I Reactor

Material Balance
(in grams)

Stream Numbers	INPUT <1>	INPUT <2>	OUTPUT <3>	OUTPUT <4>
1. Coal				
Moisture	60.0			
Coal MAF	841.1			
Ash	98.9			
2. Recirculating gas		877.3		
3. Char				
Char MAF			550.1	
Ash			98.9	
4. Tar				47.1
Light oil				18.5
Product gas				127.8
H_2O				157.6
Recirculating gas				877.3
Total	1000.0	877.3	649.0	1228.3
Total		1877.3		1877.3

Table II. Stage II Reactor

Material Balance
(in grams)

Stream Numbers	INPUT <3>	INPUT <4>	INPUT <5>	INPUT <6>	OUTPUT <7>	OUTPUT <9>	OUTPUT <10>
3. Char MAF	550.1						
Ash	98.9						
4. Product gas, tar oil and H_2O		351.0					
Recirculating gas		877.3					
5. Steam			602.7				
6. Recirculation gas				24,389.0			
7. Char MAF					164.7		
Ash					98.9		
9. Recirculating gas						25266.3	
10. Product gas							1339.1
Total	649.0	1228.3	602.7	24,389.0	263.6	25266.3	1339.1
Total				26,869.0			26,869.0

Table III. Stage I Reactor

Heat Balance

Datum Temperature 288.5°K

INPUT	Btu	Kilo Joules
1. Heating value of coal <1>	23,161	24,420
2. Sensible heat in the recirculating gas <2>	1,878	1,980
3. Total	25,039	26,400
OUTPUT		
1. Heating value of char <3>	18,064	19,046
2. Heating value of product gas <4>	3,882	4,093
3. Sensible heat in char <3>	418	441
4. Sensible heat in product gas <4>	1,140	1,202
5. Losses	1,535	1,618
6. Total	25,039	26,400

Table IV. Stage II Reactor

Heat Balance

Datum Temperature 288.5°K

INPUT		Btu	Kilo Joules
1. Heating value of char	<3>	18,064	19,046
Sensible heat in char	<3>	418	441
2. Heating value of product gas		3,882	4,093
Sensible heat in product gas	<4>	1,140	1,202
3. Sensible heat in steam	<5>	613	646
4. Sensible heat in the recirculating gas	<6>	52,217	55,056
5. Total		76,334	80,484
OUTPUT			
1. Heating value of char	<3>	3,867	4,077
sensible heat in char	<7>	312	329
2. Sensible heat in the recirculating gas	<9>	47,851	50,452
3. Heating value of the product gas	<10>	19,807	20,884
Sensible heat in product gas		2,384	2,514
4. Losses		2,113	2,228
5. Total		76,334	80,484

our experimental work is to test the bench scale set-up at UNH since conducting experiments at any of the available test facilities is expensive, and must be kept to the minimum. Thus in this first phase of our experimental work heating of the recirculation gases will be done by electric heaters. The unit is designed for eventual placement on the tower of the Georgia Tech. 400 Kw_t Solar Thermal Test Facility in Atlanta, Georgia. This facility uses large fields of sun-tracking mirrors to concentrate the sun's energy on test areas located at the focal point of the mirror field. Temperatures in the range of 1533-2366°K can be obtained at this facility.

Choice of Solar Receivers

The proposed bench scale unit will require about 10 Kw_t and an outlet gas temperature of about 1339°K. Table V lists the solar receivers which are under various stages of development. Receivers developed by Sanders Associates, Nashua, NH and Dynatherm Corporation, Cockeysville, MD, are under consideration for possible use in our system.

Sanders Associates Solar Receiver

Sanders Associates [10] has designed a ceramic matrix solar receiver. Solar energy collected by the concentrator passes through the quartz window and heats up the ceramic honeycomb which in turn transfers the heat to the working fluid <9>. A 250 Kw_t air cycle solar receiver has been tested at the Georgia Tech Advanced Components Test Facility. An output air temperature of 1339°K has been obtained with honeycomb temperatures at 1477°K. This solar receiver has been developed with funding by DOE. The use of this receiver has been assured for this project.

Dynatherm Solar Receiver

The Dynatherm Corporation has developed a

Table V

Summary of Receivers (10)

CATEGORY	COGNIZANT ORGANIZATION	PERFORMING ORGANIZATION	RECEIVER SIZE (Kw_t)	OPERATING TEMP °F	WORKING FLUID	PROJECT STATS
High Temp	JPL	Sanders	75	2000/3000	Air, HE, N_2	Conceptual Study
High Temp	JPL	General Electric	75	2000/3000	Air, HE, N_2	Conceptual Study
High Temp	DOE/JPL	MIT-Lincoln Labs	TBD	1800	Air, Others	Component Test ('79)
High Temp	EPRI	Black & Veatch	1000	1950	Air	Model Solar Test ('80)
High Temp	Sandia/L	Sanders	250	1950	Air	Model Solar Test ('79)
ADV Brayton	EPRI	Boeing	1000	1500	Air, Others	Model Solar Test
ADV Brayton	JPL	Garrett	75	1500	Air	Solar Test ('80)
Stirling	JPL	Fairchild-Stratos	75	1500	HE	Solar Test ('80)
Stirling	JPL	General Electric	75	1500	HE, NA*	Evaluation and Solar Test ('81)
ADV Rankine	SERI	Georgia Instit. Tech.	300	1100	H_2O	Performance Test ('79 - '80)
ADV Rankine	JPL	Garrett	75	1300	H_2O	Solar Test ('80)
Fuels & Chem	DOE/JPL	NRL	TBD	1700	SO_3	Advanced Development
Fuls & Chem	SERI	SERI	TBD	1200	NA**	Advanced Development
Fuels	SERI	SERI	TBD	1600	Gas+	Advanced Development
Brayton or Stirling	Dyna-Therm	Dynatherm Corp.	13.1	1950	Gas*	Solar Test ('80)

* Heat Pipe Concept

** "Thermosyphon" Concept

+ Entrained Solids in Gas (Biomass)

liquid metal heat pipe system for use as a central solar receiver (11). Its design consists of a semi-cylindrical cavity receiver with the heat pipes, mounted at the top of a tower located at the southern end of a heliostat field. The heat pipes transform the concentrated solar flux within the receiver to flux levels more compatible with efficient heat transfer to the gaseous medium to be heated. The heat pipes are made of inconel and are typically 6 cm. in diameter and 1.1 m long. They are each capable of transporting an average load of 5.3 Kw_t in the temperature range from 713°K to 1143°K. Their peak transport is 13.1 Kw_t and the highest evaporator flux is 0.34 Mw/m^2. The working fluids in the heat pipes are the alkali metals - sodium and potassium.

Economics of Commercial Solar Gasification Plants

In our earlier paper (3), an economic analysis for a plant capacity of 7.08 x 10^6 m^3 (250 MM SCF) per day of SNG has been presented. This capacity was recommended for conventional coal gasification plants planned in the 1960's. However, no such plant of this capacity has as yet been built in the U.S.A.

The cost estimates for solar hybrid units indicated that by 1995 the cost of gas (SNG) from such units would be competitive with conventional gasifiers. This analysis was based on the projected cost of heliostats which depends critically on the number of heliostats manufactured. Once the use of solar energy for the production of electricity, industrial process heat, or fuels and chemicals is well established, the number of heliostats produced will increase substantially and thereby the cost of heliostats will decrease. Further, there is no realistic estimated cost for solar receivers at present as the number manufactured has been insignificant. As in the case of heliostats the cost of solar receivers will also decrease with the increase in the number of units produced.

A solar hybrid unit of 7.08 x 10^6 m^3 (250 MM SCF) per day of SNG capacity requires 1750 Mw_t of solar energy. To supply this amount of solar energy approximately 3275 acres of space is necessary to install the required number of heliostats. In view of the extensive area needed, a study of the effect of the size of the hybrid plant on the cost of SNG has been made. The 'six-tenths' rule is used to compute the investment for plants of different capacities and the labor cost is assumed to follow the 'four-tenths' rule. When the cost of SNG is plotted against capacity of the plant (Fig. 2) it is seen that the cost of gas increases slowly with a decrease in capacity from 7.08 x 10^6 m^3 (250 MM SCF) to 1.13 x 10^6 m^3 (40 MM SCF) per day and then exponentially when the capacity is further reduced below 0.566 x 10^6 m^3 (20 MM SCF) per day. The most suitable size thus appears to be about 1.42 x 10^6 m^3 (50 MM SCF) per day of SNG for a solar hybrid plant. The requirement of solar energy for this size gasification plant including boilers, utilities, etc. is estimated to be 350 Mw_t.

Typical industrial process heat (IPH) solar thermal plants planned for operation by the year 2000 are expected to have an average capacity of 500 MM Btu/hour or 147 Mw_t (12). The land area required for housing the required number of heliostats is estimated to be about 275 acres. It is also expected that 100 such plants will be in operation by the year 2000. If a solar hybrid coal gasification plant is to be matched with a typical industrial process heat (IPH) solar thermal plant that will be in operation in the year 2000, the capacity of the hybrid unit must be only about 25 MM SCF/day. Hence, for a capacity of about 1.42 x 10^6 m^3 (50 MM SCF) per day of SNG production two such typical IPH units will be required.

Conclusions

Gas recirculation hybrid solar gasification systems have the potential for producing synthesis gas on a continuous basis. Preliminary heat and material balances indicate that no major experimental problems should be encountered in our bench scale experiments.

The reactors to be used in this study are made of stainless steel and not of quartz nor do they have any windows to clean. Therefore the data obtained can be directly used for scaling up to pilot plant size, an intermediate step towards the final goal of commercialization of this technology. By the year 2000, typical IPH solar thermal plants expected to be in operation will have a capacity of 500 MM Btu/hour or 147 Mw_t. Two such units will be needed to supply enough heat to operate one 1.42 x 10^6 m^3 (50 MM SCF) per day solar gasification plant.

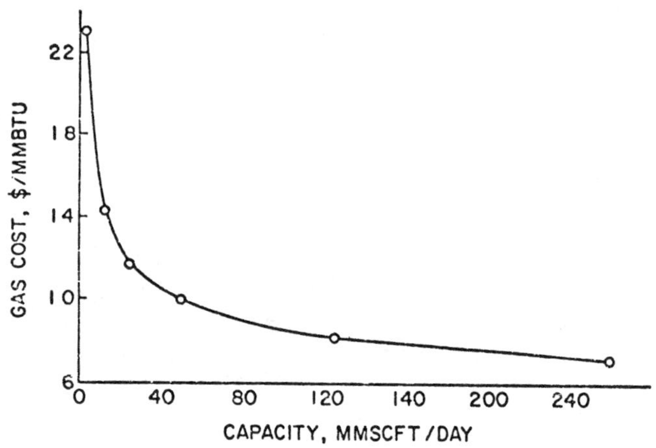

Fig. 2 Gas Cost Dependence on Plant Capacity

The realization that solar thermal energy can play a vital role in the production of synthetic fuels is only now beginning to receive the attention it deserves. The use of solar thermal energy for the production of fuels and chemicals directs the thrust of solar energy to our most critical national need of convenient energy storage. One can envision a complex of solar "mines", located in regions of optimum solar insolation (most probably the American Southwest) where solar energy will be collected and converted to sun fuels, then shipped by a pipeline to user centers throughout the nation. Availability of this stored solar energy to users will be independent of cloud cover, the day-night cycle, local insolation, or the vagaries of local weather.

LITERATURE CITED

1. Gregg, D.W., W.R. Aiman, H.H. Otsuki, and C.B. Thursness, "Solar Coal Gasification", Solar Energy 24, 313, 1980.

2. W.R. Aiman and D.W. Gregg, "Gasfication of Coal with Solar Energy," paper presented at the 88th National Meeting of AIChE, Philadelphia, PA, June 1980. [Lawrence Livermore Laboratory Report UCRL-84458, June 1980.]

3. Lakshmanan, S., F.K. Manasse, and V.K. Mathur, "Production of Fuels from High Temperature Solar Thermal Systems - Economic Analysis," AIChE Symposium Series 198 Vol. 76, 156, 1980.

4. "Alternate Fuels from High Temperature Solar Thermal Systems", UNH Report on DOE Grant No. DE-A C02-79 ET 21067, May 1980.

5. Copeland, R.J., "Rough Cost Estimates of Solar-Thermal Coal or Biomass Derived Fuels," SERI-TP-35-279, June 1979.

6. Antal, M.J., "Synthesis Gas Production from Organic Wastes by Pyrolysis/Steam Reforming," presented at IGT Conference in Energy from Biomass and Wastes, Washington, DC, Aug. 1978.

7. Hesp, W.R. and P.L. Waters, "Thermal cracking of tars and volatile matter from coal carbonization," I&EC Prod. Res. Develop. 9 (2) 194, 1970.

8. Colaluca, M.A., M.A. Paisley and K. Mahajan, "The Tri-Gas Gasification Process," Chem. Eng. Progr. 75 (6) 33, 1979.

9. Institute of Gas Technology; Preparation of a Conversion Systems Technical Databook. U.S. Energy Res. Develop. Admin. Report No. FE-1730-21.

10. Kudirka, A.A. and L.P. Leibowitz, "Advanced Solar Thermal Receiver Technology", paper presented at the 18th Aerospace Science Meeting, Pasadena, California. Jan. 1980.

11. Bienert W. and D. Wolf, "Liquid Metal Heat Pipes for the Central Solar Receiver", 13th IEC-EC Conference, San Diego, CA, Aug. 1978.

12. Private communication, "Solar Thermal Program Multi Year Plan", 1980.

SOLAR PRODUCTION OF INDUSTRIAL PROCESS STEAM IN CHEMICAL PROCESS INDUSTRIES

The potential for the utilization of solar energy for the production of industrial process steam in chemical processing industries is discussed briefly. To advance the technical and economic feasibility of solar-thermal systems applications, NL Industries of Newberry Springs, California, a DOE funded demonstration project, has been chosen as a case study.

The NL Industries solar system consists of 950 square meters of Del single-axis, tracking, parabolic-trough, concentrating collectors. It was designed to produce a portion of the 420°-530°K steam utilized in a drying operation to reduce the moisture content of hectorite ore from 10 percent to 4 percent. (Hectorite is a hydrous magnesium silicate which, when refined, is of significant commercial interest because of its applications in various chemical and food processes.) It is estimated that implementation of this solar system could result in an annual savings of 3.545 billion KJ (3.360 billion BTUs), of the equivalent of 90 cubic meters (600 barrels) of oil and a net reduction of 1,860 kilograms (4100 pounds) of air pollutants annually. The technical, economic and institutional issues encountered in the course of this project are also discussed.

Finally, the impact of the commercialization of solar energy applications in chemical processing industries is evaluated.

S. SUNDARAM
and
B. G. ELDRIDGE

Jacobs Engineering Group, Inc.
251 South Lake Avenue
Pasadena, California 91101

The industrial sector is the largest consumer of energy in the United States, consuming about 40 percent of the nation's gross energy demand of 77 quads (quad = 10^{15} KJ) annually. It is estimated that from 50 to 70 percent of this demand is for industrial process heat (IPH). At least 27 percent of required industrial process heat (4.5×10^{15} KJ) applications require temperatures of below 560° K; this temperature ceiling allows the use of commercially-available concentrating collectors.

It is interesting to note that the largest energy users in the United States, in order of their use, are: (1) the minerals and metals industries; (2) the chemical industry; (3) the petroleum industry; and (4) the transportation industry. The work reported herein was designed to aid the development of the solar production of industrial process steam. Specifically, the objective of the project was to establish technically, economically and environmentally acceptable methods for the generation of process steam for mineral, chemical and food processsing industries.

SOLAR TECHNOLOGY

The use of solar energy as a viable and clean energy has aroused widespread public interest and government action. Since solar energy can be used for water heating, space heating and cooling for buildings, agricultural and industrial process heat and electric power generation, solar collectors are classified according to their working temperature and geometry.

Flat Plate Collectors

The simplest collector geometry is the flat plate. Flat plate collectors are well suited for water heating and space heating for buildings. Since they do not employ optical concentration, they collect both diffused and direct solar radiation and can supply building heat output even under hazy or cloudy conditions. Flat plate collectors are economically competitive with electric heaters. Their economical working temperature ranges from 330° to 350° K.

Concentrating Collectors

Cylindrical, parabolic-trough collectors can achieve optical concentration ratios exceeding 60. (A concentration ratio is defined as the ratio of the light intensity on the absorber to unconcentrated sunlight intensity.) Since cylindrical, parabolic-trough collectors cover a temperature range of 360° - 590° K, they are widely used for industrial process heat applications. These collectors can be arranged in series; in parallel, or in combinations of both, depending on the system working temperature and

0065-8812-81-4534-0210-$2.00

capacity. The collector efficiencies can be improved by altering tracking mechanisms and collector orientation.

Parabolic dish collectors are capable of generating the highest temperatures of any of the concentrating collectors and are the most optically efficient systems among distributed systems. Two-axis tracking in parabolic dish collectors virtually eliminates cosine loss, since the aperture is always normal to the direct beam radiation. Concentration ratios are as high as 3000 for parabolic dish collectors. These collectors can be used for the production of steam, for electric power generation, or for both. The technology for this collector needs further research, development and demonstration, however.

Other concentrating collectors, such as fixed-compound parabolic, fixed mirror slat, vee trough and fixed hemispherical dish, are similar to the above-mentioned collectors except for slight modifications in the collector modules.

Power Tower

A point-focus central receiver system, often called a "power tower", concentrates reflected sunlight on an elevated, heat absorbing receiver. This absorbed energy is used to produce electricity, process steam, or both. The large field of mirrors (also known as heliostats) employs two-axis tracking. The tower is placed near a central location in the heliostat field.

APPLICATIONS--DESIGN PHILOSOPHY

Based on experience, Jacobs-Del Solar Systems has established several criteria for the selection of cost-effective IPH applications. These are summarized below:

1. A conservation program at the prospective industrial facility is essential.
2. Land/roof area must be available for collector installation. (Land is preferable, since ground installations are the most cost-effective.)
3. Conditions must permit locating the solar collector field close to the plant to minimize thermal losses and to simplify the solar system interface with the existing process.
4. Maximum utilization of solar energy with little or no thermal storage must be an objective.
5. A careful evaluation must be made of significant conditions: location of plant, solar irradiation and ambient air quality.
6. Industrial-grade hardware must be utilized for solar systems.

SIMULATION MODEL[3]

Jacobs-Del has developed a computer simulation model to evaluate our approach to designing systems for the solar production of industrial process heat. This model, the Solar Thermal System Simulation (STSS) program, will allow us to select a cost-effective system from candidate solar systems.

The three major components of the STSS are a collector field performance program, a transient thermal losses in pipes and equipment program and an economics program. The field performance program determines the amount of usable energy which an array can collect under varying conditions. The second STSS program estimates thermal losses in pipes and storage tanks during a 24-hour day. The STSS economics program considers factors such as capital investment, life expectancy, operating and maintenance costs, taxes, fuel savings, etc. in the process of evaluating the cost of an IPH system on the basis of dollar per million BTU and dollar per square foot of collector.

CASE STUDY--NL INDUSTRIES, NEWBERRY SPRINGS, CALIFORNIA[4]

Process

The NL Industries facility at Newberry Springs processes hectorite ore. Hectorite ore is a hydrous magnesium silicate which, when refined, is of significant commercial interest because of its applications in various chemical and food processes. The refinement process includes a drying operation which reduces the moisture content of the product to 4 percent. Medium-pressure steam of 1,100 KP is utilized in this process. The implementation of a solar energy system to produce a portion of the 420° - 560° K steam required would provide a useful demonstration of the feasibility of the solar production of industrial process steam, and the facility's Mojave Desert location makes it an ideal site for solar studies. For these reasons, in response to a U.S. Department of Energy Request for Proposal, we explored various solar energy systems for NLI to determine which would be the most cost-effective on a $/KJ basis.

Concept Design Analysis

Our original proposal was for a base system consisting of 1,425 square meters of Del collectors installed on a 15° tilt angle. At this time we anticipated that a new Del collector with a two-meter aperture and an evacuated receiver tube would be designed and ready for production in January of 1980. (Unfortunately, the development program for this collector slipped on to two years.) Several alternatives to the base system were also offered. The vital statistics on each of these alternatives are summarized for comparison in Table 1. To conform to DOE budgetary limitations, however, it became necessary to modify the suggested base system. The new design reduced the array to 950 square meters of collectors installed on a 7.5° tilt angle. A supplementary analysis was then conducted to evaluate variations of this new, smaller system. These eight alternatives are compared in Table 2.

System Selection

An in-depth study proved that the pump-down vacuum tube system required further development before implementation/use. Table 2 indicates that the Solar Kinetics collector might provide the most cost-effective system on a $/KJ basis for collectors with standard receiver tubes. However, we consider the Del collector superior--certainly so for the desert location--due to its glass reflective surface. For this reason, the Del collector with the standard receiver tube (Alternative #1, Table 2) was considered the best option and was selected as the basis for the final design.

System Details

The solar system designed for NL Industries consists, as previously stated, of 950 square meters of Del, single-axis, tracking, parabolic-trough, concentrating collectors. The collectors are divided into four groups of 160 collectors each, positioned on a 7.5° tilt, facing South, to maximize the annual collector of energy by reducing end and cosine losses. Eight collectors comprise the delta temperature strings; the drive string includes 32 collectors. All collectors and piping are supported on concrete piers. Figure 1 shows a solar steam flow diagram.

The solar energy collection loop is a closed-circuit piping system, pressurized with nitrogen. A circulating pump circulates the heat transfer fluid from the solar array to a steam generator, then back to the solar array. Because of its high heat transfer characteristics, water is to be used as the heat transfer fluid. The system is designed to produce hot water at a temperature of 500° K and at a pressure of 2,900 KPa (420 psig).

Solar energy will not be stored at high temperatures at the NL Industries facility. When possible, it will be utilized, as collected, to generate steam. When there is insufficient solar energy for the generation of steam, the low-grade solar energy will be collected and used to heat water in a 90 cubic meter process hot water storage tank.

The current system at Newberry Springs employs a boiler, fired with natural gas, to produce steam; fuel oil is used when the natural gas is interrupted. The steam is fed directly to drum dryers. Condensate from the drum dryers is returned to a deareator, from which boiler feed water is supplied to the boiler.

A flue stack economizer on the existing 9,525 Kg/hr. boiler will provide waste heat energy to the new solar system. In addition, a new heat exchanger will recover waste heat energy to increase the efficiency of the solar energy system. The economizer will preheat the process hot water and the feed water for the boiler.

Most operations at the Newberry plant run 24-hours a day; the steam drum drying facility, however, is a 12-hour/day, 363/day/year operation. Waste heat, which will always be available from the economizer, will be utilized to preheat the solar collection system in the early morning hours. This will allow early steam generation with solar energy and will avoid a warm-up period, the length of which would ordinarily be determined by ambient temperature and the availability of solar radiation. The waste energy will be used, also, to replace the energy in the solar system which will be lost during hours of cloud cover or at night.

PROBLEMS ENCOUNTERED--AND THEIR SOLUTIONS

Technical

Two potential problems have been anticipated and remedied. First, freezing could occur during the winter months; waste heat from the economizer, or hot water from the storage tank will provide freeze protection. Second, collector maintenance and cleaning could be complicated by fine dust blown across the site from an ore storage area west of the system; this storage area has been relocated.

Economic and Institutional[5]

After the oil embargo and the continued increase in world crude oil prices, the national laboratories and industries involved in solar energy began to examine the economic and institutional issues which impeded the commercialization of solar installations for the agricultural, commercial and industrial sectors. The institutional issues which would foster commercialization include federal and state energy tax credits, federal investment tax credits, property tax exemptions, pollution offset credits, low-interest loans and income tax rebates on fossil fuel saved. Some of these issues, such as state and federal energy tax credits and federal investment tax credits, have been enacted. But other issues, including property tax exemptions and pollution offset credits, are yet to be resolved in some states, including the State of California.

Property Tax

Solar property tax exemptions are the most popular financial incentive. 28 states currently offer real property tax exemptions for solar systems. Since the solar devices become permanent fixtures, the real property is subject to property taxes when it is attached to an existing plant or building. In addition to the solar devices incorporated, the professional services rendered during the design and construction are taxable realty by definition. Normally, in states favoring solar property tax exemptions, the pipes, pumps, storage tanks, instrumentation, etc., are included in the property tax exemption. The conventional, back-up systems are excluded.

The property tax issue can be resolved using one of the following approaches:

1. Solar systems are assessed at no more than comparable, conventional, fossil-fuel systems.
2. Solar systems are either partially or entirely exempted from property taxation.
3. Rights are transferred to local governments to enact measures which exempt solar systems from property taxes.

28 states have enacted one of these three approaches. In the State of California, the property tax exemption bill was passed in June, 1978. However, this tax exemption apparently does not apply to solar energy process heat applications.

Pollution Offset Credit

Federal and State environmental policy has been faced with two, conflicting goals: the legislation imperative to clean-up the nation's air and the desire of cities to attract new industries. Although the conflict is still not completely resolved, the EPA ruled in 1976 that industrial expansion would be allowed only if it resulted in net reduction in emissions. Thus, an air pollution trade-off policy was established. It permitted the reduction of emissions from existing sources to offset emmissions from new plants. The use of solar energy systems could accelerate both industrial and economic growth without adding air pollutants. The EPA should consider expanding the pollution offset credit specifically to include solar energy systems. Solar systems would then be attractive to industrial and commercial sectors. A pollution offset credit based on dollars/square foot of collector would be helpful.

CONCLUSION

Commercialization of solar energy for industrial purposes is an attractive goal in light of current concern over air pollution and the availability and price of fossil fuels. Based on our experience, nearly half of the total system cost goes for collector sub-systems. This is mainly due to the production of very small quantities of collectors. In order to compete economically with conventional fuels, a combination of matured technology and mass production of collector sub-systems is a must.

Limited governmental action in the form of energy and investment tax credits has encouraged interest in the utilization of this energy source. Further legislation in the areas of property tax exemptions and pollution offset credits is essential, however.

Most solar systems are not economically competitive today with fossil fuel systems. However, industries should relax their pay-back period up to ten years for viable energy options. Also, the government should provide low-interest loans for industries adopting energy alternatives such as solar. Thus, solar energy can contribute its share to industries without the degradation of air, water and land.

LITERATURE CITED

1. D.M. Considine, "Solar Technology," p. 6-6 to 6-134, Energy Technology Handbook, McGraw-Hill, New York (1977).

2. F. Kreith and J.F. Kreider, Principles of Solar Engineering, McGraw-Hill, New York (1978).

3. S. Sundaram, C.F. Roos and B.G. Eldridge "Computer-Aided Solar Thermal Systems Analyses for Industrial Process Heat (IPH) Applications," DOE/SERI 2nd Annual Systems Simulation and Economic Analysis Conference, San Diego, California, January 1980.

4. B.G. Eldridge, "The Solar Production of Industrial Process Steam," 2nd Annual Solar Industrial Process Heat Applications Conference, Oakland, California, November 1979.

5. B.G. Eldridge and S. Sundaram, "Institutiional Constraints in Commercializing Solar Energy for Process Heat Applications," 45th International Conference on Assessment Administration, Las Vegas, Nevada, October 1979.

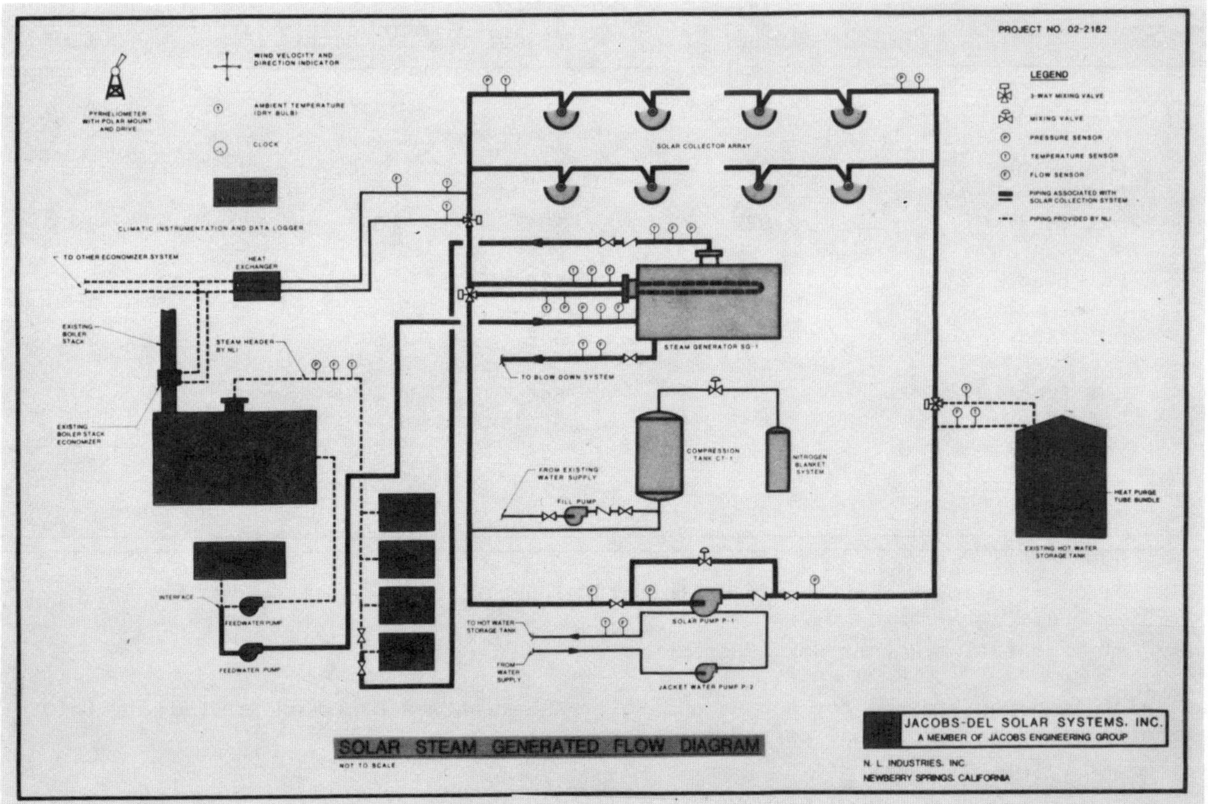

Figure 1.

Table 1. Comparison of Alternative Solar Systems

Orientation: North/South Tilt Angle 15°

Alter-native	Area of Array (Sq. Meter)	Usable Solar (10^9 J)	Est. Con-struction Cost ($000)	Unit Costs $/Sq. Meter	Unit Costs $/$10^9$ J
1	1,425	5,016	803	560	160
2	1,425	5,016	781	550	156
3	2,000	10,539	886	440	84
4	2,100	11,458	873	420	76
5	1,600	9,166	764	475	83

Table 2. Comparison of Alternative, New, Smaller Solar Systems

Alter-native	Collector	Tilt Angle	Usable Solar* (10^9 J)	Est. Cost ($000)	Unit Costs $/Sq. Meter	Unit Costs $/$10^9$ J
1	Del (Std)	7.5°	3,168	638	670	201
2	Del (Vac)	7.5°	4,408	708	720	161
3	Del (Std)	0	2,872	594	625	207
4	Del (Vac)	0	4,000	664	700	166
5	Del (Std)	0**	2,729	576	600	211
6	Del (Vac)	0**	3,800	646	680	170
7	Solar Kin.+	0**	2,728	553	580	190
8	Acurex+	0**	2,592	635	670	245

* Fuel Savings = (Usable solar energy/boiler efficiency)

** East/West Orientation; other North/South

\+ The calculated performance for Acurex and Solar Kinetics was based on preliminary information only and not on actual performance data.

THE POTENTIAL FOR SUPPLYING SOLAR THERMAL ENERGY TO INDUSTRIAL UNIT OPERATIONS

E. KENNETH MAY
Solar Energy Research Institute
1617 Cole Boulevard
Golden, Colorado 80401

The major near-term impact of using solar energy to supply process heat will most likely occur in industries operating processes at temperatures below 370°C (700°F). Applying the "end-use matching" concept effectively ranks industrial unit operations as solar energy applications. Drying industrial materials by direct exposure to solar radiation is a unique opportunity to provide solar industrial heat economically and on a large scale.

Using high-temperature water (HTW) for process heating can be more economical than using steam. Solar systems are more easily retrofit to liquid heating systems, and transporting heat as HTW could solve problems of land availability that may hinder widespread application of solar technology.

The growth of the solar industry is greatly influenced by the cost of alternative fuels and by government policies. Solar systems can be installed quickly and could have an immediate impact in reducing fossil fuel use. Increased involvement by engineers in industry and firms engaged in the design, construction, and use of conventional energy facilities will accelerate the attainment of this goal.

Major reports published by Battelle (1) and InterTechnology Corporation (ITC) (2) contain surveys of industrial energy usage based on standard industrial classifications. An analysis of typical processes employed in each industry determined the quantity of fuels used for industrial process heat (IPH) and the temperature requirements of these processes. According to the American Petroleum Institute, of the 21 quads* of energy consumed by industry in 1976 (3), 68% is estimated to have been used for process heating, distributed in the temperature ranges illustrated in Figure 1 (2).**

Just over 50% of industrial process heat is used at temperatures above 593°C (1100°F). The industries operating in this temperature range are few, chiefly primary metals; stone; clay and glass products; and chemicals. The number of facilities operating such high-temperature, energy-intensive processes is also small. For instance, 1973 there were only 43 integrated iron and steel plants in the United States.

*A quad is 10^{15} Btu (1.055×10^{18}J).

**Recent ongoing studies at the Solar Energy Research Institute (SERI) suggest that the Battelle and ITC reports overestimate the quantities of IPH consumed at high temperatures. Actual consumption seems to be distributed more toward the low-temperture ranges. The ITC data suffice, however, to illustrate the points made in this paper.

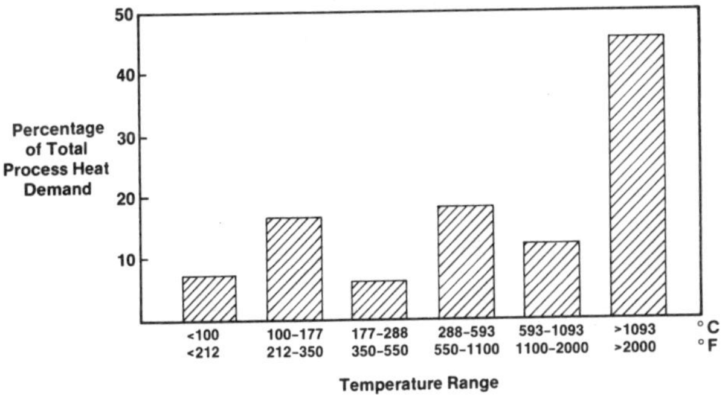

Source: Reference [1], 1977.

Figure 1. Distribution of Industrial Process Heat Demand by Temperature Range (1974)

Over 95% of process heat in the range of 288° to 593°C (550° to 1100°F) is consumed by the petroleum refining industry (4). Currently, 297 domestic oil refineries are operating (5). This range also encompasses the temperatures at which steam is generally produced for electrical generation.

0065-8812-81-4029-0210-$2.00

Process heat used below 288°C (550°F) comprises about 28% of the total used by industry. The number of plants operating low-temperature industrial processes is very large. For instance, there are approximately 2900 textile and 2500 dairy establishments. Such plants are widely scattered and are much less energy-intensive than plants operating high-temperature processes. This paper postulates that plants operating low-temperature processes offer the most favorable market for the near-term application of solar thermal technology to provide industrial process heat. By matching process temperature requirements as closely as possible with solar collector operating temperatures, industrial unit operations are broadly ranked according to the ease with which solar radiation can be economically utilized as a source of heat.

SOLAR TECHNOLOGIES

Low-Temperature Applications

There are a few difficulties in using solar technology to provide heat to processes operating below 371°C. Flat-plate collectors or solar ponds can deliver temperatures up to about 82°C (180°F). Evacuated-tube collectors operate up to about 149°C (300°F), and line-focusing collectors--for example, parabolic troughs--operate with reasonable efficiency up to about 288°C (550°F). Central-receiver technologies utilizing the "power-tower" concept allow high solar concentration ratios to be achieved. Consequently, a temperature of 371°C is easily within their operating range, and such a temperature allows the use of conventional organic heat transfer fluids as energy transport media.

Most of the countless number of plants operating industrial processes below 371°C use relatively little energy. Such facilities typically burn premium fuels such as natural gas or distillate, and, consequently, are particularly vunerable to large increases in fuel costs or curtailments. Coal is rarely considered an option due to its inherent environmental problems. Apart from the largest heating requirements, low-temperature IPH needs could probably be met using the less sophisticated, fairly well developed solar technologies, such as parabolic troughs, flat-plate collectors, or solar ponds. Some industries do have very large demands for low-quality process heat. The corn processing industry uses considerable quantities of hot water and low pressure steam. Several mining operations--for example, the processing of uranium ore--use fuel primarily to heat water. About 20% of the total energy supplied to a refinery is used to provide process heat at temperatures below 371°C (6). The energy intensity of refineries is such that even this fraction represents nearly 0.8 quad or about 1% of the total U.S. energy consumption. This energy is primarily used for the reboiling of distillation columns, where considerations of pressure drop and temperature control are much less critical than for higher temperature catalytic processes and where conventional heat exchangers and heat transfer fluids can be used. The cogeneration of electrical power and process heat is always an option for large users of low-temperature heat if the dual uses are suitably matched. The size of the solar collector field for large-scale applications could favor central-receiver technology. For large fields, compared to distributed-receiver technology, the increased thermal efficiency and reduced piping costs and electrical inputs of a central-receiver system could prove economically advantageous.

The large number of plants operating low-temperature industrial processes is an advantage in that the characteristics of some will be favorable to the early application of solar technology; e.g., high fuel costs, location in a rural area with high levels of solar radiation, and progressive management. On the other hand, this diversity can be a disadvantage because small plants may not have skilled technical and engineering work forces, intense competition can discourage risks not associated with the development of new products, investment capital may be lacking, and personnel may not have access to the required information.

Intermediate-Temperature Applications

The practicality of using solar technology to generate IPH in the range of 288° to 593°C can be determined by a study of the predominant user: the petroleum refining industry. Total energy consumption of this industry is about 11% of the energy content of the crude processed (7). This amount is equivalent to about 1.5 million barrels/day of crude oil or nearly 4% of total U.S. energy consumption.

The technical characteristics of refinery process heat use are apparent from a study of the schematic of the naphtha reforming unit illustrated in Figure 2. To minimize power inputs to the compressor and for process reasons, pressure drop through the recycle system is kept as low as possible. This requires specialized design of heaters, reactors, and exchangers. Refinery units are crowded into the processing area to reduce piping runs and to allow fluids to be transported between units with minimal temperature and pressure drops. Fired heaters operating on fluid fuels allow extremely high rates of heat transfer. Such fuels have very consistent properties, thus permitting the accurate temperature control so important to catalytic processes (6).

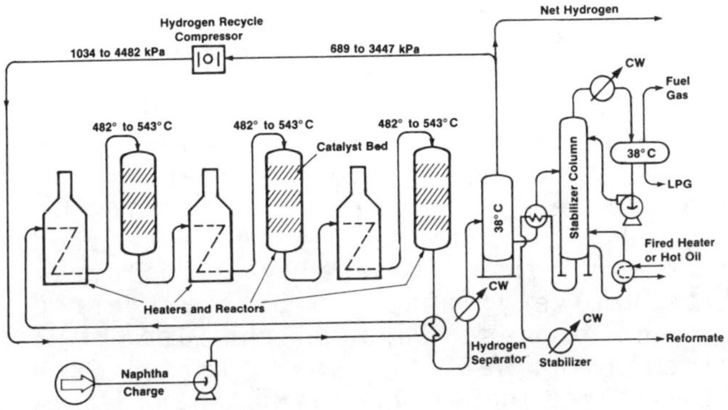

Figure 2. Catalytic Reforming

At the present state of development of solar thermal technologies, it is difficult to conceive of a solar system operating with the performance of fired heaters and fitting into the refinery processing environment. A fired heater is compact and allows very high rates of heat transfer with minimal holdup, low pressure drop, and accurate temperature control. A solar hot-air system would have to operate well in excess of 593°C (1100°F) to minimize heat exchanger surface area. Space limitations could hinder the integration of a solar system into the refinery processing area. A liquid heat exchange system raises the problem of the availability of heat transfer fluids. Operations above about 371°C (700°F) are outside the range of commercially available organic heat transfer liquids and would probably involve the use of molten salts or metals. The mechanical specifications of a hydrocarbon/molten-salt exchanger would be extremely rigorous and molten salts or metals, particularly for safety reasons, are not desirable media to be pumped around an oil refinery. In summary, the application of solar thermal technology to petroleum refinery processes operating above 371°C presents numerous technical difficulties. Other industrial applications in the intermediate temperature range pose similar technical difficulties. Significant displacement of fossil fuels at these temperatures will require considerable progress in solar technology and process development research.

High-Temperature Applications

While solar furnaces can generate extremely high temperatures, there are great technical and economic obstacles to applying solar thermal technology to industrial processes requiring temperatures of 593°C (1100°F) or higher. The adaption of solar technology to industrial processes in the iron and steel industry, for instance, would require the use of unconventional, high-cost materials to collect and transport solar energy at these temperatures. Significant problems involving creep, thermal fatigue, and their interaction need to be overcome (8). Major breakthroughs in materials technology will be required before the application of solar energy at temperatures greater than 593°C is a pratical proposition.

Coal presents serious competition to any possible use of solar energy for high-temperature processes. It is a very economical form of energy and most high-temperature processes can readily use coal as a source of fuel. In iron production, there is no easy substitute for the reducing properties of carbon. High-temperature processes are heated by direct contact of the flue gas with the process material--a very efficient heating method. Equipment design can result in a high degree of air and charge preheat. The large size of high-temperature operating facilities greatly enhances the economics of coal distribution, storage, and handling, and can justify the expense of environmental controls when necessary. Sometimes coal can be burnt with minimal environmental effects; for example, in lime production the product efficiently absorbs fly ash and sulfur. In short, for the near term at least, it is probable that solar technology will find most applications for process heating at temperatures above 593°C on a specialized basis only.

LOW-TEMPERATURE UNIT OPERATIONS

End-Use Matching

One aspect of the work carried out at the Solar Energy Research Institute (SERI) has been an industrial case studies program aimed at determining the near-term economic and technical feasibility of applying solar thermal technology to industrial processes (9). Most studies have been performed in response to inquiries from industry. It is apparent that some of the best prospects for the application of solar systems were overlooked by the large-scale surveys of industrial energy use. There is a need to disseminate information to firms regarding conservation measures and alternative energy sources. Future industrial decision making concerning energy use is likely to involve much more than a selection of the appropriate size of steam generator. A simplified means of evaluating the potential of solar technology could be extremely helpful in this decision process. One means of simplifying this evaluation is to focus on the end use of energy delivered to industrial unit operations.

A major requirement in considering the application of any solar technology is to closely match the solar collector outlet temperature with the required industrial process temperature. This gives rise to the concept of "end-use matching." The performance of all solar thermal systems is affected by thermal losses to an extent much greater than conventional heat generating equipment. As a result, their efficiencies depend quite strongly on the collector operating temperature. It is also generally true that less sophisticated collectors, such as flat-plate collectors operating in their efficient temperature range, can deliver energy more cheaply than more complicated collectors, such as parabolic troughs. The end-use matching approach allows some important conclusions to be made regarding the potential of solar thermal technology for providing industrial process heat.

The unit operations of drying, evaporation, crystallization, distillation, baking, and extraction, as well as general heating for applications such as pasteurization, reactor heating, cleaning, dissolving, and sterilization, are universal throughout industry and are most often carried out at temperatures below 371°C. For process temperatures below about 204°C (400°F), steam is the preferred heat-transport medium, even though about 7.5% of total process heat usage occurs at temperatures below 100°C, mainly to heat water. To provide heat to processes operating in the 204° to 371°C range, direct-fired heaters are generally employed. Alternatively, heat can be transferred to the process through the use of a circulating heat transfer fluid. Such fluids are generally used when it is not possible to pump the process fluid through the coils of a heater, when it is desired to distribute heat to various locations from a centrally located heater, or for safety reasons.

Industrial processes for which direct solar heating of the process medium in the collector field is possible afford the opportunity to match closely collector and process operating temperatures. Applications to produce hot water for cleaning, sterilization, and dissolving are particularly favorable. Obviously, because the performance and cost of solar collectors depend strongly on operating temperature, the lowest-temperature applications are most favorable. In fact, before the installation of any solar system is contemplated, a thorough analysis of the industrial process is required to determine the operating temperature actually required by the process. This temperature is not necessarily the historical operating temperature.

If possible freezing of the collector fluid presents problems, the above processes and the operations of evaporation, crystallization, distillation, and pasteurization can be carried out by the transfer of heat from a nonfreezing heat transfer fluid circulating through the solar collectors. Indirect heating increases equipment costs and lowers collector performance because of higher collector operating temperatures.

Drying, baking, and curing involve heating air. While drying is theoretically possible at almost any temperature, baking and curing are generally carried out at fairly high temperatures, typically about 204°C. Present industrial practice generally is to use direct gas-fired heating. Such high-temperature, hot air processes are less favorably matched to solar thermal technologies than the operations previously described because of the temperature levels, the high efficiency use of conventional fuels, and the poor heat transport characteristics of air reduce the efficiency of hot-air solar collectors. Also, moving compressible fluids through great lengths of collector piping consumes considerable quantities of electrical energy. The technical solution for solar hot-air applications at temperatures too high to allow direct heating is to use a liquid heat transport medium and a liquid/air exchanger. Consequently, baking and curing operations, compared to other processes conducted at the same temperature are probably the least favorable solar applications. However, pursuing the end-use matching approach leads to some important insights regarding the drying process and the use of steam as a heat transport medium.

Applications of Solar Energy to the Drying Process

The drying process offers a unique opportunity for the application of solar technology. Drying is a major consumer of industrial process heat and many drying processes operate at temperatures below 288°C. All manner of materials, such as food, lumber, painted and printed materials, and metal ores, are dried. The energy content of biomass and wet coal is enhanced considerably by drying. Much drying is carried out at quite low temperatures to avoid degradation of the material. For instance, plastic films must be dried at temperatures below 93°C (200°F), and lumber-drying kilns typically operate at 38° to 82°C (100° to 180°F). A considerable amount of drying is also carried out at temperatures up to about 649°C (1200°F), although, from a thermodynamic point of view, drying is possible at almost any temperature.

Drying at high temperatures reduces the residence time of the process material, and this reduces the size and capital cost of the drying equipment. Most drying takes place near the wet-bulb temperature and this tends to prevent thermal degradation of the dried product. Often, some loss of quality--for example, in drying animal feed--can be tolerated. High-temperature drying is more efficient thermally and requires less electrical power than drying at low temperatures because of the reduced quantities of air involved. For instance, drying an agricultural product using air at 121°C (250°F), as opposed to air at 316°C (600°F) assuming the same relative humidity at the dryer outlet), requires 1.6 times the thermal input and 4.8 times the air flow rate. It is unlikely that low-temperature solar energy could be supplied so cheaply as to compete with efficient high temperature processes. However, several factors enhance the potential of the solar drying process. First, the drying of materials, particularly at high temperatures, entails high gas velocities, and carry-over of solid material can cause particulate pollution problems. Drying at low temperatures requires longer contact times and hence reduced gas velocities, and the pollution potential is lowered. Second, the drying of many materials at high temperatures--for example, food products such as milk--depends on the direct use of clean-burning and increasingly expensive natural gas. The alternatives to natural gas are probably fuel oil or coal (if available and environmentally acceptable) for generating steam to run an indirect, air-heating process for drying. Such a change brings the drying process into a temperature range that is readily attainable through the use of solar collectors. Finally, and most importantly, if solar technology is a viable alternative, problems of low thermal efficiency and high power inputs, which are inherent to low-temperature drying processes, can be overcome by the direct exposure of the process material to solar radiation. One can envision a transparent glazing over a slowly moving conveyor. Besides providing highly efficient use of solar radiation, which could replace several times the equivalent number of Btu obtained from conventional sources of fuel, such a system could be constructed very cheaply. The rate of air flow need be sufficient only to control the humidity level of the outlet gas. Gas velocities would be so low as to present little possibility of entrainment. Materials for which the possibility of contamination is not a problem, such as coal, animal feed, biomass prior to combustion, and corn, are prime candidates for such drying. Coal is particularly appropriate because it is a good absorber of solar radiation.

High-Temperature Water as an Alternative to Steam for Heat Transport

Steam has become the universal, low-temperature, heat transport medium even though it is often generated at temperatures far in excess of the required process temperature, and even though few processes other than for example, steam-stripping or humidification necessitate its use. Package steam boilers are available "off the shelf" to meet a wide variety of process heat requirements. They are compact, rapid in response, and extremely reliable. Pumping power requirements are low, and the feed water pump runs at a low temperature, though at high head. The condensation process allows large quantities of heat to be transferred at a constant temperature and low mass flow rate--a particular advantage for direct process heating. However, a body of trade literature (10) advocates that the package steam boiler is a relic of the era of low fuel costs. Process heating using high-temperature water (HTW) is feasible up to about 204°C for replacing saturated steam as a heat transport medium. Above this temperature, heat transfer fluids with lower vapor pressure can be used. Liquid systems, compared to steam systems, allow more accurate process temperature control. Energy and money are saved as a result of reductions in water makeup, treatment, and blowdown. The large thermal mass of liquid systems reduces fluctuations in heater output, thus improving heater efficiency and reducing wear. Water is a convenient medium for thermal energy storage. Fewer fittings and less corrosion potential also reduce maintenance costs compared to a steam system. The use of HTW to transport heat over large distances appears to offer significant economic benefits compared to any other medium (11). Thermal transport of energy as HTW could solve problems of land availability that could hinder widespread application of solar technologies for industrial process heating.

A switch from steam to more energy-efficient, liquid-based systems involves a trade-off between the present cost of capital and future energy savings. The liquid system requires greater heat transfer area on the process side, larger return lines, and a high-capacity circulating pump. If fluid fuels are being replaced, as energy costs rise much faster than inflation, the balance should continue to shift in favor of capital investments. Another trade-off common to energy conservation measures and to the use of solar systems concerns the relative cost of energy used as heat versus the cost of increased use of electrical power. Energy-saving measures such as HTW systems, forced-draft furnaces, or vapor recompression cycles greatly complicate a process and require additional electrical inputs. However, because the price of electricity in most parts of the United States is based largely on coal, nuclear, and hydroelectric sources, the cost of electricity is likely to increase at a lower rate than the costs of either oil or natural gas. From 1972 to August 1979, for example, the cost of industrial electricity rose 150%, compared to 499% for the wellhead price of natural gas and 364% for the price of refinery fuel oil (12).

The Advantages of Liquid-Based Solar Systems

Although the relative merits of HTW and steam systems for the provision of IPH can be argued, there is no doubt that conditions favoring the industrial installation of liquid heating systems would drastically improve the economics of solar-derived energy. It is relatively easy to retrofit liquid-based heating systems. Figure 3 illustrates one commonly used solar steam system. To prevent boiling in the receiver tubes (the effects of which are unknown), high-pressure water is pumped through the collectors. Steam is generated by flashing across a pressure-reducing valve and discharging into a vapor/liquid separator. Pressure drop through the system is high and, as a result, electrical parasitic power consumption is high. The solar collectors must operate at

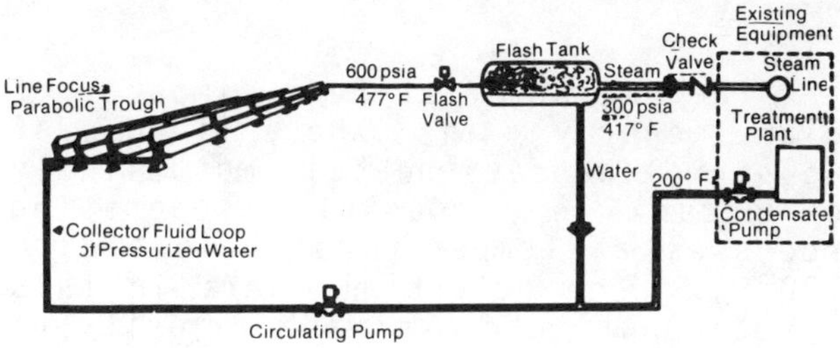

Figure 3. Flash Steam Generating System

temperatures much higher than the steam temperature, at considerable detriment to their efficiency. Figure 4 illustrates how a solar system could be integrated into a HTW pasteurization process operated by a dairy. An expansion tank is already a feature of the existing liquid heating system. Solar collectors are simply added as an additional loop.

Liquid-based solar systems allow a close match between technology and process temperature, which is consistent with the end-use matching approach. Thus, flat-plate collectors or solar ponds, in particular instances, could be used to provide heat at temperatures below about 82°C. Using such technologies to produce hot water is probably a more economic solar option than adapting completely to the existing, conventional energy delivery system, which would probably entail the generation of steam at a much higher temperature.

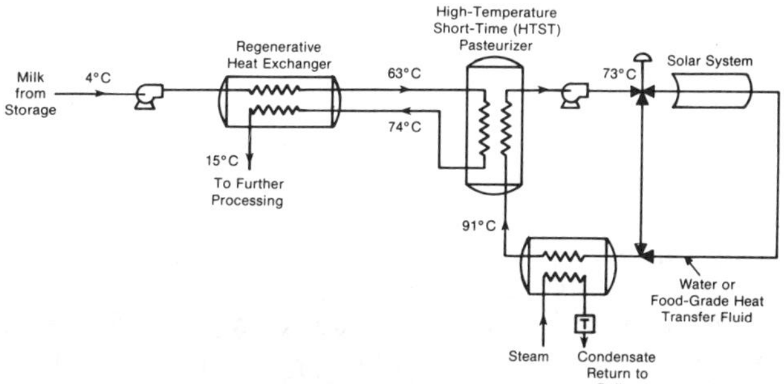

Figure 4. Solar-Heated Milk Pasteurization Process

SUMMARY

The major near-term impact of solar energy in the field of industrial process heat is most likely to occur in industries operating processes at temperatures below 371°C (700°F). The potential users in this market constitute the vast majority of all industrial establishments.

Through a procedure of matching the operating temperature of a solar collector as closely as possible with the industrial process temperature, unit operations have been discussed with respect to their suitability for solar energy applications. Processes such as the heating of water for cleaning purposes, which allow the direct solar heating of the process medium, are highly favorable solar applications. Processes such as evaporation and crystallization, which require indirect heating, rank next. The least suitable process applications involve the heating of air to high temperatures. However, the drying process is a unique opportunity to provide solar IPH economically and on a large scale. Solar drying, as envisioned here, would involve a revolution in currently employed industrial drying practices. It would require a return to age-old concepts of direct solar drying and the application of low-technology solar components.

Another processing consideration emphasized in this paper is the advantage of HTW systems for replacing steam as a heat transport medium. The increased use of HTW systems in industry considerably enhances prospects for the retrofit of solar IPH systems.

Other factors, such as government policy and the cost of alternative fuels, will greatly influence the growth of the solar industry. While no one doubts the long-term potential of renewable energy resources, it is a fact that solar technology can have an immediate impact on the energy supplies. The lead time for installing solar systems is very short compared to other energy-producing technologies. This potential will be realized, however, only through the active participation of engineers in industry who have gained a working knowledge of solar technology. Perhaps what is needed more than anything else is a sense of commitment and a willingness to take some risks. The potential for solar technology to provide industrial process heat, and the benefits that this could bring, will only be realized if this commitment is forthcoming from government and from industries currently involved in the design, construction, and use of conventional energy facilities.

ACKNOWLEDGMENTS

This work was supported by the U.S. Department of Energy under the Division of Solar Thermal Energy Systems. The material presented in this paper is derived from work conducted at the Solar Energy Research Institute to determine the characteristics of industrial energy use and the potential for solar energy to provide industrial process heat.

LITERATURE CITED

(1) Battelle Columbus Laboratories, "Survey of the Applications of Solar-Thermal Energy Systems to Industrial Process Heat," 3 vols., Battelle Columbus Labs, Columbus, OH (1977).

(2) InterTechnology Corporation, Analysis of the Economic Potential of Solar Thermal Energy to Provide Industrial Process Heat, 3 vols., C00/2829-76/1, InterTech Corp., Warrenton, VA (1977).

(3) American Petroleum Institute, "Basic petroleum data book," API, Washington, DC (1977).

(4) Ketels, P. A.; Reeve, H. R., Market Characterization of Solar Industrial Process Heat Applications, PR-52-212, Solar Energy Research Institute, Golden, CO (1979).

(5) Cantrell, Ailleen, Oil and Gas Journal, 78 (12), (1980).

(6) May, E. Kenneth, Solar Energy and the Oil Refining Industry, TR-733-562, Solar Energy Research Institute, Golden, CO (1980).

(7) Gordian Associates, Inc., An Energy Conservation Target for Industry SIC 29," 2 vols., prepared for the Federal Energy Office, Gordian Associates, New York, NY (1976).

(8) Murr, L. E., Solar Materials Science, Academic Press, New York, NY (1980).

(9) Hooker, D. W.; May, E. K.; West, R. E, Industrial Process Heat Case Studies, TR-733-323, Solar Energy Research Institute, Golden, CO (1980).

(10) Teller, William H.; Diskant, William; Malfitani, Louis, "Conserving fuel by heating with hot water instead of steam," Chemical Engineering, 185-190 (1976).

(11) Hausz, Walter, Combined Thermal Energy Storage and Transport for Utility Applications," EPRI EM-1175, Electric Power Research Institute, Palo Alto, CA (1979).

(12) Farrar, Gerald L, Oil and Gas Journal, 78 (1), (1980).

COMMERCIAL-SCALE PROCESS DESIGN FOR THIN-FILM SOLAR CELLS

A preliminary process design for the manufacture of thin-film cadmium sulfide/copper sulfide solar cells is outlined. The total product cost is determined from raw materials, utilities, labor, and capital using engineering economic procedures of the chemical processing industry. Results are given for manufacturing facilities with annual throughput of 10, 100 and 1000 megawatts. Batch and continuous modes of operation are compared and it is shown that thin-film solar cells can be manufactured for less than $0.50 per watt (50 $/m²) in a continuous production facility. Capital investment to meet this low price will be between $263 and $347 million, an investment roughly equivalent to a world-scale methanol plant.

T. W. F. RUSSELL
B. N. BARON
and
R. E. ROCHELEAU

Institute of Energy Conversion
Department of Chemical Engineering
University of Delaware
Newark, Delaware 19711

INTRODUCTION

Solar energy is both abundant and diffuse, delivering to North American latitudes a peak energy flux $\sim 1 kW/m^2$. Photovoltaic solar cells are semiconductor devices which directly convert sunlight into electricity.

Increasingly since 1973, the development of low-cost high-efficiency solar cells has been a goal shared by both government and industry. Today, solar cells based on crystalline silicon with efficiencies in the range of 10% to 15% are commercially available at $7.50 to $11 per peak watt. The acceptance of solar electricity generation in the marketplace requires further reductions in cost. The U. S. Department of Energy, has set a goal of $0.70 per watt by 1986. Since large areas of solar cells must be deployed in order to generate significant amounts of energy, the cost and utilization of raw materials is a key factor in reaching this goal. One gains an appreciation for the magnitudes involved by considering the areas of cells required. For example, a processing plant designed for 1000 MW of peak solar electric capacity produces 10^7 square meters (3.6 square miles) of 10% efficient solar cells per year. Solar cells using thin semiconductor films instead of thick crystals have long been recognized as meeting the requirement of low materials cost. Thin-film solar cells based on cadmium sulfide/copper sulfide have been a subject of research since the photovoltaic behavior of this materials system was first reported nearly twenty years ago. A major effort was undertaken during the space program (1) in order to exploit the low weight and radiation resistance of CdS cells. Efficiencies were low, rarely exceeding 5%, and yields were poor compared to silicon. Since 1973, research has been directed towards improving the efficiency of CdS thin-film cells. The basic physics (2) and materials science (3) were primary concerns in the initial stages. Considerable effort was devoted to device design (4). In 1978, a laboratory-scale 1 cm^2 CdS/Cu_2S cell having an efficiency of 9% was reported (5). However, in order to meet the requirement for thin-film cells to be produced at a commercial scale, process design and detailed economic analysis are necessary.

Much of the analysis of commercial-scale solar cell manufacturing to date has followed the historical approach of the semiconductor integrated circuit industry. Considerable attention has been given to the device, its operation and materials. However, solar cells at the commercial scale are very much a chemical product. The procedures of the chemical process industry are not only applicable, but necessary.

The steps to develop a commercial-scale design from bench-scale laboratory experiments are shown in Figure 1. The batch experi-

0065-8812-81-4025-0210-$2.00

mentation in solar cell work concerns itself, in the initial stages, with developing the necessary basic physics and material science. After the fundamentals have been established, considerable effort needs to be devoted to device design. In carrying out the bench-scale experimentation, limited attention is usually devoted to the apparatus used to build the devices. In order to construct equipment which will serve at a commercial scale, it is essential that the performance of each equipment piece be modeled and the model checked by experiment. The preliminary process design is carried out in three stages: A study of each unit operation leads to mathematical descriptions of each piece of apparatus which are used to design commercial-scale equipment. The major equipment is conceptually linked together to develop the process or system needed to produce solar cells at a commercial rate. The third part of process design is determination of capital and operating costs for the system. Cost data are used to develop optimal designs and to determine whether the commercial-scale process is economically feasible.

Sometimes a commercial-scale unit is built without further developmental research. In spite of the risks, large profits can be realized if the commercial-scale unit is started up far enough ahead of the competition to dominate the market. However, more than half the time that this approach is followed, the commercial-scale unit has start-up difficulties and becomes an expensive pilot plant or unit operations laboratory. A more conservative route to the commercial scale involves one, or sometimes two, additional stages. Unit operations are carried out on equipment five to ten times the size of bench-scale equipment in order to solve first order scale-up problems. This equipment frequently has different configurations than the bench-scale equipment, e.g., it may be designed to run continuously rather than in the batch mode. A pilot plant operation will be set up in conjunction with, or following, unit operations experimentation. In certain cases, the unit operations experiments provide sufficient information to build the commercial-scale unit without pilot plant operation. A benefit of pilot plant operation is a supply of product for market development. It has the additional advantages of testing the total system at an expense level considerably below that of the commercial-scale plant. The main disadvantage is the time needed to design, construct and operate the pilot plant. At least two years would be required for most thin-film solar cell pilot plants.

In this paper, we shall outline process and manufacturing costs for thin-film solar cells at the commercial scale. The basis of this paper is application of the chemical process industry's unit operations approach to the 9% cell laboratory process.

DESCRIPTION OF THE CdS/Cu_2S CELLS

Figure 2 shows a cross-sectional view of the cell. The cell is comprised of five functional layers:

Opaque Contact

The substrate/ohmic contact is electroformed copper foil electroplated with 0.3-0.5 micron of zinc. Besides providing ohmic contact, the substrate layer is an efficient reflector of useable light for cells operating in the front-wall-reflection mode (6).

Collector/Converter

The collector layer is cadmium sulfide formed by sublimation in vacuum of cadmium sulfide powder at 900° to 1100^oC and condensation on the heated (180°-250^oC) substrate. The laboratory process and fundamental materials properties of the cadmium sulfide are described elsewhere (7).

Absorber/Generator

The absorber layer is cuprous sulfide formed by topochemical ion exchange to a thickness of 0.1-0.3 micron. Cuprous sulfide for the 9% cell of (5) was formed by solution reaction (8). Since higher efficiency (9) and yield are expected from solid-state reaction (10), the following discussion will be limited to the solid-state reaction.

Transparent Contact

The transparent contact consists of close-spaced metallization in a parallel stripe geometry (11).

Encapsulation/AR

The final functional layer is an antireflection coating and a protective glass barrier layer to provide encapsulation.

PROCESS DESIGN

For the purpose of process design, deposition of the cadmium sulfide and copper sulfide active semiconductor layers, contacts, and encapsulation are treated as separate unit operations, Figure 3.

Each of the operations used in the laboratory for fabrication of the 9% cell have been analyzed. Detailed models have been developed for the vacuum deposition processes. From this analysis, preliminary process designs and cost estimates for manufacture of cadmium sulfide solar cells in plants with annual throughputs of 10^5, 10^6 and 10^7 square meters active area have been completed. Process designs and cost estimates have been completed for both batch and continuous modes of operation.

Batch operation is typified by the processing of many small substrate pieces. The batch process may be highly automated. Continuous processing denotes a reel-to-reel operation, a technology widely used in the coating industry (12). Reliable costs for reel-to-reel units are available from equipment manufacturers. Truly continuous processing a strip from initial substrate to finished cell would replace reel-to-reel operation only if economic benefits could be proven.

Fabrication procedures for the two modes of operation are nearly identical. However, labor requirements, materials handling equipment, and key equipment pieces are not the same. Key unit operations in the manufacture of cadmium sulfide cells are discussed below.

SUBSTRATE PREPARATION

The substrate used in the laboratory, Zn-plated copper foil, is available in the quantities needed for a commercial-scale plant (13). The foil must be cleaned before CdS deposition. Standard electrocleaning and acid pickle equipment, chemicals and procedures are available from suppliers.

The two modes of production differ mainly due to the handling equipment. The continuously-processed substrate can be unwound under tension, pulled through baths and rewound at a takeup location. Batch processing requires preliminary process steps to cut the substrate and mount it in suitable carrier frames.

CdS Deposition

Cadmium sulfide is deposited on the substrate by thermal evaporation. Evaporation rate, substrate temperature, and background pressure are controlled to obtain desired film properties. Appropriate ranges of each variable have been defined in the laboratory process. The source temperature is controlled to provide deposition rates between 1.5 and 2.0 microns per minute to a thickness of 10 microns. Photovoltaic grade films are obtained with substrate temperature between 200°C and 240°C (7). Vacuum systems with power supplies and controls for both batch and continuous coating are available from a number of firms. A continuous reel-to-reel vacuum deposition unit has been designed to deposit films of uniform thickness with greater than 60% material utilization.

The batch deposition process design is based on an estimated 45 minutes per cycle for unloading, loading and pumpdown in addition to 6 to 10 minutes of deposition time. Continuous deposition onto a roll of material effectively eliminates turnaround time. Process control is facilitated with continuous processing because transient changes are eliminated. In addition, batch evaporation requires rigid substrate mounting to prevent curling, and is more susceptible to outgassing problems.

Cuprous Sulfide

The process design assumes the so-called "dry-process". In the dry process, cuprous chloride is evaporated onto the cadmium sulfide film to a thickness of 0.1 to 0.3 microns. Reaction takes place during a 3 to 5 minute heat treatment in a non-oxidizing atmosphere. A solvent or water rinse to remove cadmium chloride and drying complete the process.

The major equipment expense is the vacuum deposition unit. The source is similar to that used for CdS but the temperature requirements of the substrate during deposition are less stringent. The reaction can be carried out in the evaporator immediately following deposition. The technical advantages of continuous versus batch operation for cuprous sulfide formation are the same as for cadmium sulfide deposition.

Transparent Front Contact Application

The present laboratory method of electron beam evaporation of gold through an aperature mask is neither technologically nor economically promising for commercial-scale processes.

The transparent front contact is composed of three identifiable components: grid, terminal buss insulator, and terminal buss. The commercial-scale design is based on 95% transparent grids deposited by electrolytic or printing techniques. Equipment costs are based on state-of-the-art printing techniques. Materials costs are based on non-noble metal grids with the option of 0.1 micron of gold or carbon to ensure ohmic contact. The terminal buss insulator is a thermoplastic or inorganic paste printed around the borders of the cell and cured in an inert atmosphere. This layer insulates the terminal buss bars from the back contact. The terminal buss is printed over the insulator, using a suitable conductive paste.

Screen, contact and rotogravure printers as well as photoresist processes are available for batch and continuous operations. The main advantage of continuous processing is decreased handling equipment costs.

Encapsulation

Technology development for integral encapsulation of front-wall cadmium sulfide cells at optimum efficiency is currently at the laboratory stage of fundamental physics and materials science. Process design and cost estimates are preliminary and, therefore, conservative.

A two-layer encapsulant is assumed. The first layer is an evaporated or sputtered glass 0.1 to 1. micron thick. This coating, formed on the cuprous sulfide, is a hermetic seal and anti-reflection layer. A protective layer of 1/16 inch tempered glass is then bonded to the cell with polyvinyl butyral.

COST ANALYSIS

Total product cost is calculated using the cost of raw materials, utilities, direct operating labor, and capital investment. The procedure for computing costs is the method used by the chemical process industry (14). Raw materials are calculated directly from mass balances on the various equipment pieces and process obtained from suppliers. A summary of raw material requirements and cost on a per watt basis is shown in Table 1 for a 1000 megawatt plant. Cost per watt is the same for other plant sizes. These requirements are based on a cell efficiency of 10% with an insolution of one kW/m^2. The active cell area, including grids, is 80% of total substrate area. The remaining 20% is attributed to buss tabs or waste. Cadmium sulfide requirements are based on 75% of the evaporated material adhering to the substrate. No recycle of unused material is assumed. Other evaporation steps assume the same utilization efficiency. All other material usages are based on the mass balances for processing 100% of the total substrate area, 1.25×10^5 m^2/yr, 1.25×10^6 m^2/yr and 1.25×10^7 m^2/yr for the 10, 100 and 1000 megawatt plants respectively. Ranges of costs are used in Table 1 when the materials may be substituted, e.g., gold, or the specification remains open, e.g., plate glass. No change in materials utilization with throughput is assumed. An optional 5 mil brass substrate is coated for the batch process because it may be needed to reduce curling. Major equipment items were sized from a detailed flow sheet of the process. Costs were obtained directly from suppliers whenever possible. Where uncertainties exist, a range of costs are given.

Purchased equipment costs for 10^7 m^2/yr (1000 MW) batch and continuous plants are shown in Table 2. Also included in Table 2 are estimates of operating labor. As expected, the costs associated with the steps using vacuum are many times greater than any of the other costs. Non-vacuum equipment makes up only 2.5% of the total equipment cost. The maximum dimensions of the vacuum chambers were 3 meters in diameter by approximately 2.5 meters long. The maximum strip width was two meters. With these chamber dimensions and the aforemetioned range of deposition rates the design strip speed was 32 to 43 cm/min (1-1.5 ft/min). Continuous vacuum coating units, capable of coating two meter wide webs at speeds greater than 100 ft/min are commercially available. The same chamber length was used for the batch process since larger substrate would be difficult to handle. While larger vacuum deposition units might be designed to trade unit cost against proliferation, the vacuum deposition units given in Table 2 are accurately costed items, available from several competing manufacturers. Operating labor for each operation shown in Table 2 was estimated based on the complexity and number of equipment pieces.

A summary of capital and labor requirements is shown in Table 3. The fixed capital investment for a CdS/Cu_2S solar cell plant was estimated to be 445% of uninstalled equipmask is neither technologically nor economically promising for commercial-scale processes.

The transparent front contact is composed of three identifiable components: grid, terminal buss insulator, and terminal buss. The commercial-scale design is based on 95% transparent grids deposited by electrolytic or printing techniques. Equipment costs are based on state-of-the-art printing techniques. Materials costs are based on non-noble metal grids with the option of 0.1 micron of gold or carbon to ensure ohmic contact. The terminal buss insulator is a thermoplatic or inorganic paste printed around the borders of the cell and cured in an inert atmosphere. This layer insulates the terminal buss bars from the back contact. The terminal buss is printed over the insulator, using a suitable conductive paste.

Screen, contact and rotogravure printers as well as photoresist processes are available for batch and continuous operations. The main advantage of continuous processing is decreased handling equipment costs.

Encapsulation

Technology development for integral encapsulation of front-wall cadmium sulfide cells at optimum efficiency is currently at the laboratory stage of fundamental physics and materials science. Process design and cost estimates are preliminary and, therefore, conservative.

A two-layer encapsulant is assumed. The first layer is an evaporated or sputtered glass 0.1 to 1. micron thick. This coating, formed on the cuprous sulfide, is a hermetic seal and anti-reflection layer. A protective layer of 1/16 inch tempered glass is then bonded to the cell with polyvinyl butyral.

COST ANALYSIS

Total product cost is calculated using the cost of raw materials, utilities, direct operating labor, and capital investment. The procedure for computing costs is the method used by the chemical process industry (14). Raw materials are calculated directly from mass balances on the various equipment pieces and process obtained from suppliers. A summary of raw material requirements and cost on a per watt basis is shown in Table 1 for a 1000 megawatt plant. Cost per watt is the same for other plant sizes. These requirements are based on a cell efficiency of 10% with an insolution of one kW/m^2. The active cell area, including grids, is 80% of total substrate area. The remaining 20% is attributed to buss tabs or waste. Cadmium sulfide requirements are based on 75% of the evaporated material adhering to the substrate. No recycle of unused material is assumed. Other evaporation steps assume the same utilization efficiency. All other material usages are based on the mass balances for processing 100% of the total substrate area, 1.25×10^5 m^2/yr, 1.25×10^6 m^2/yr and 1.25×10^7 m^2/yr for the 10, 100 and 1000 megawatt plants repectively. Ranges of costs are used in Table 1 when the materials may be substituted, e.g., gold, or the specification remains open, e.g., plate glass. No change in materials utilization with throughput is assumed. An optional 5 mil brass substrate is coated for the batch process because it may be needed to reduce curling. Major equipment items were sized from a detailed flow sheet of the process. Costs were obtained directly from suppliers whenever possible. Where uncertainties exist, a range of costs are given.

Purchased equipment costs for 10^7 m^2/yr (100 MW) batch and continuous plants are shown in Table 2. Also included in Table 2 are estimates of operating labor. As expected, the costs associated with the steps using vacuum are many times greater than any of the other costs. Non-vacuum equipment makes up only 2.5% of the total equipment cost. The maximum dimensions of the vacuum chambers were 3 meters in diameter by approximately 2.5 meters long. The maximum strip width was two meters. With these chamber dimensions and the aforemetioned range of deposition rates the design strip speed was 32 to 43 cm/min (1-1.5 ft/min). Continuous vacuum coating units, capable of coating two meter wide webs at speeds greater than 100 ft/min are commercially available. The same chamber length was used for the batch process since larger substrate would be difficult to handle. While larger vacuum deposition units might be designed to trade unit cost against proliferation, the vacuum deposition units given in Table 2 are accurately costed items, available from several competing manufacturers. Oeparting labor for each operation shown in Table 2 was estimated based on the complexity and number of equipment pieces.

A summary of capital and labor requirements is shown in Table 3. The fixed capital investment for a CdS/Cu_2S solar cell plant was estimated to be 445% of uninstalled equipment cost using the methods of Peters and Timmerhaus (14). Operating labor requirements were based on 24-hour year-round operation with three shifts requiring 1.33 man-years per operator-year of labor to keep the plant running.

Total product costs for the 10, 100 and 1000 megawatt facilities are shown in Table 4. Utility requirements were determined from equipment specifications. Ranges are shown, indicating uncertainties in the present state-of-the-art process design for CdS/Cu_2S thin-film solar cells.

CONCLUSIONS

Procedures used by the chemical process industries for computing total product costs are applicable to thin-film solar cell manufacture.

Cadmium sulfide-copper sulfide solar cells can be manufactured for less than $0.50 per watt with continuous (reel-to-reel) processing using equipment operating at moderate speeds. Batch processing is approximately 70% more costly. Capital investment requirements range from $12 to $18 million for a 10 MW continuous unit, and $300 to $400 million for a 1000 MW continuous unit.

Raw materials are the highest single manufacturing cost, accounting for up to 38% of the total product cost. No other single manufacturing cost accounts for more than 15% of total product cost.

LITERATURE CITED

1. L. R. Shiozawa, et. al. Final Report on U. S. Air Force Contract AF-33(615)-5224 (1969).

2. A. Rothwarf, Proc. Intl. Workshop on CdS and Other Abrupt Heterojunctions, Univ. of Delaware AER-75-15858, p. 9 (1975).

3. J. D. Meakin, Photovoltaic Materials Device Workshop, SERI/TP-49-185 p. 11 (1979).

4. A. Rothwarf & A. M. Barnett, IEEE Trans. Electron Devices, ED-25 (4), 381 (1977).

5. A. M. Barnett et al, Thirteenth IEEE Photovoltaic Specialists Conference Proceedings p. 419 (1978).

6. J. A. Bragagnolo, ibid p. 412.

7. R. B. Hall & J. D. Meakin, Thin Solid Films 63, p. 203 (1979).

8. B. N. Baron et al, Thirteenth IEEE Photovoltaic Specialists Conference Proceedings p. 406 (1978).

9. A. Rothwarf, Twelfth IEEE Photovoltaic Specialists Conference Proceedings p. 488 (1976).

10. T. S. teVelde, Energy Conversion 14, 111 (1974).

11. W. E. Devaney et al, IEEE Trans. Electron Devices ED-26 (3), p. 205 (1979).

12. D. M. Mattox, See (2) , p. 494.

13. Gould Inc., Thin Foil Division, McConnelsville, OH; Yates Industries, Bordentown, NJ.

14. M. Peters and J. Timmerhaus, Plant Design & Economics for Chemical Engineering, 2nd Ed. McGraw Hill, N. Y. (1968).

FIGURE 1

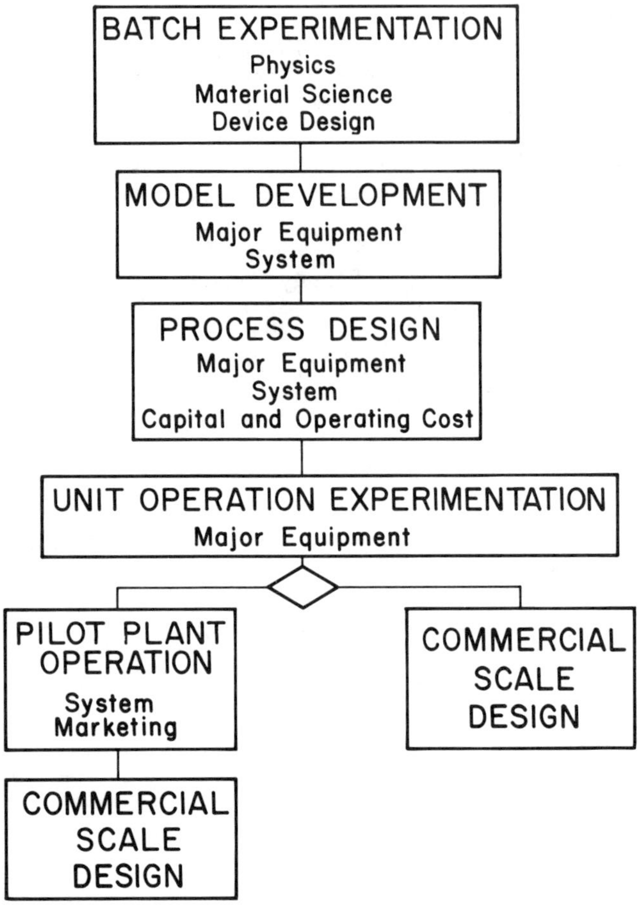

FIGURE 2

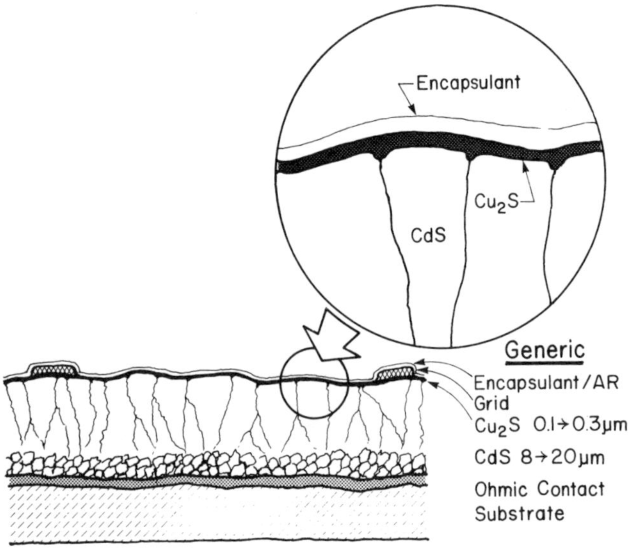

FIGURE 3

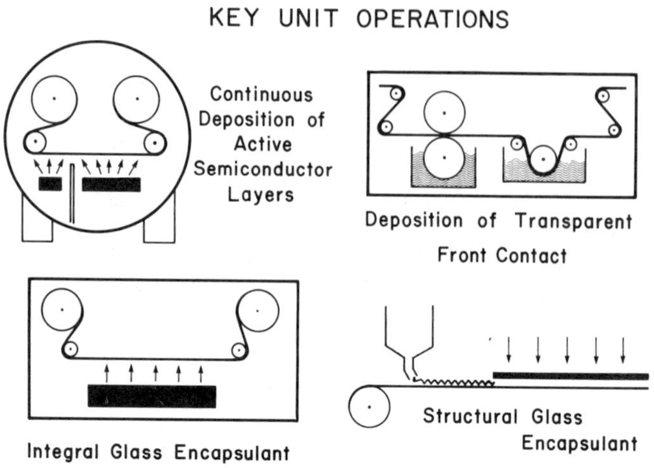

TABLE 1

SUMMARY OF RAW MATERIAL COSTS FOR 10^7 m^2/yr SOLAR CELL PLANT (1000 MW)

RAW MATERIAL	QUANTITY (lb/yr)	COST ($/watt)	PER CENT OF COST
Plate Glass Encapsulant	1.25 x 10^7 m^2/yr	.039 - .075	40 - 45
Gold Grid (1000Å, 10% Area)	5.3 x 10^3	0 - .034	0 - 20
25 Micron Brass Substrate (125 Micron Brass Substrate)	6 x 10^6 30 x 10^6	.029 .064	30 - 17
Cadmium Sulfide (10 Micron, 60% Util)	1.8 x 10^6	.020	20 - 12
Buss Insulator (TPA-85)	22 x 10^3	.001	1
Integral Glass Encapsulant	370 x 10^3	.004	4 - 2
PVB Binder	5.5 x 10^6	.003	3 - 2
Cu (Grid Line, Buss)	172 x 10^3	.0005	1
CuCl	24 x 10^3	.0002	1
Miscellaneous		.001	1
Total		0.10 - 0.17	

TABLE 2

UNINSTALLED PURCHASED EQUIPMENT FOR 10^7 m^2/yr SOLAR CELL PLANTS (1000 MW)

UNIT OPERATION	BATCH		CONTINUOUS		MINIMUM NUMBER OF PROCESSING UNITS
	COST*	LABOR+	COST*	LABOR+	
Substrate Preparation	805 - 1,130	11	455 - 620	5	1
CdS Evaporation	47,000 - 70,500	40	19,000 - 25,000	10	40 - 50
CuCl Evaporation/Reaction	47,605 - 59,490	45	19,580 - 25,730	13	40 - 50
Grid Print/Plate	1,090 - 1,390	8	890 - 1,140	4	1
Integral Encapsulant Evap.	47,000 - 58,750	40	19,000 - 25,000	10	40 - 50
Plate Glass Installation	110 - 140	7	190 - 240	7	1
Misc. Scrubbers & Precipitators	80 - 100	1	80 - 100	1	
TOTALS	143,690 - 191,500	150	59,195 - 77,830	50	

* All Costs in Thousands of Dollars (1979)

+ Estimated Labor in Operators Per Shift

TABLE 3

CAPITAL AND LABOR REQUIREMENTS

PLANT SIZE	CAPITAL Millions \$ (1979)		LABOR (Person-Years)	
	BATCH	CONTINUOUS	BATCH	CONTINUOUS
10^5 m^2/yr (10 MW)	12.3 - 17.9	10.4 - 15.3	68	56
10^6 m^2/yr (100 MW)	80 - 107	34.3 - 45.4	168	80
10^7 m^2/yr (1000 MW)	636 - 854	263 - 347	600	200

TABLE 4

BREAKDOWN OF TOTAL PRODUCT COST FOR SOLAR CELL MANUFACTURING FACILITIES (All Costs in 1979 Dollars Per Watt)

	COST FOR 10^5 m^2/yr (10 MW)		COST FOR 10^6 m^2/yr (100 MW)		COST FOR 10^7 m^2/yr (1000 MW)	
	BATCH	CONTINUOUS	BATCH	CONTINUOUS	BATCH	CONTINUOUS
I. TOTAL MANUFACTURING COST	.80 - 1.14	.68 - .94	.45 - .68	.26 - .39	.37 - .57	.22 - .34
A. Capital	.15 - .23	.13 - .19	.10 - .14	.04 - .06	.08 - .11	.03 - .05
B. Raw Materials	.10 - .20	.10 - .17	.10 - .20	.10 - .17	.10 - .20	.10 - .17
C. Utilities	.08 - .14	.06 - .10	.06 - .10	.03 - .06	.06 - .10	.03 - .05
D. Production Labor	.25 - .30	.21 - .25	.10 - .12	.05	.07 - .08	.03
E. Overhead & Misc.	.22 - .27	.18 - .23	.09 - .12	.04 - .05	.06 - .08	.03 - .04
II. NON-PRODUCTION COSTS	.24 - .50	.19 - .41	.10 - .25	.06 - .14	.07 - .20	.04 - .12
III. TOTAL PRODUCT COST	1.04 - 1.64	.87 - 1.35	.55 - .93	.32 - .53	.44 - .77	.26 - .46

TEMPERATURE EFFECTS ON THE KINETICS OF THE ANAEROBIC DIGESTION OF CROP RESIDUES TO METHANE

Crop materials may be converted to methane by the process of anaerobic digestion. Since the reactions are quite slow, process parameters must be investigated to help optimize methane yield. This study investigates the effect of temperature on the kinetics of the anaerobic digestion of cornstalks to methane. A preliminary process design and process economics are also discussed.

E. C. CLAUSEN
and
A. H. SHAH
Department of Chemical Engineering
Tennessee Technological University
Cookeville, Tennessee 38501

Anaerobic digestion is an old and proven process used extensively for treating domestic sewage sludge until about 1930. The process has more recently received attention as a method of producing energy in the form of methane gas. Many raw materials have been used in the process including food and industrial wastes (5, 15), feedlot wastes (9), garbage (14), algae (7) and agricultural residues (4). Agricultural residues in the form of cornstalks, oat hulls, etc. are especially promising raw materials for anaerobic digestion because they are quite abundant and have minimal agricultural uses.

The anaerobic digestion process operates at atmospheric pressure and at moderate temperatures. It occurs in the absence of oxygen in three basic steps. First, solid organic material such as cornstalks or grasses is solubilized. The soluble organics formed are then metabolized by bacteria to organic acids and alcohols. Methane bacteria then convert these acids and alcohols to methane and carbon dioxide .

Previously completed kinetic studies involving anaerobic digestion have shown the overall reaction to be quite slow, with retention times of 15 to 25 days quite common (4). These long reaction times and the low bulk density of the raw material make large reaction vessels necessary. This, in turn, translates into high capital costs for the reaction vessels. Indeed, preliminary process design calculations have shown that the cost of reaction vessels constitute a major portion of the total capital investment (4). Thus, future research work involving anaerobic digestion should concentrate on improving the rates of reaction in order to decrease capital costs for the process.

PARAMETERS AFFECTING ANAEROBIC DIGESTION

Many physical parameters may be investigated to help increase the overall rate of methane production in the anaerobic digestion process, such as pH, temperature, nutrients concentration, agitation rate, and particle size of the feed material. Another approach is to optimize the individual steps of the anaerobic digestion process. A study concerning separating the three stages of the anaerobic digestion process was carried out by Clausen and Gaddy (3). This study showed an increased overall rate of reaction and, hence, greater methane production in the separate stage study. However, the process economics indicated that the methane produced in a three step process was more expensive.

0065-8812-81-4520-0210-$2.00

The price of methane in the separate stage study was about $0.23 per thousand liters in comparison to $0.12 per thousand liters in the single-stage study. This difference was mainly due to the additional equipment needed in the separate stage study.

Physical parameters such as reaction temperature, pH, agitation rate and nutrients concentration should be investigated, since optimizing these parameters should not increase capital costs by introducing additional pieces of equipment but may lead to an increase in overall reaction rate. A proper pH balance for the many pH-sensitive simultaneous reactions is necessary so that the formation of methane or any reaction intermediate is not inhibited. The optimum reaction temperature must be sought to maximize the reaction rate without causing death to the highly temperature sensitive organisms. All necessary nutrients for growth must be identified and must be present in the proper quantities. Optimum levels of feed particle size and reactor agitation rate must also be found as a function of power input to the system.

Many of these parameters have been studied by various investigators for the digestion of sewage sludge. However, optimum values may vary with the raw material fed and should, of course, be studied with process economics kept in mind.

THE EFFECTS OF REACTION TEMPERATURE ON DIGESTION

As previously mentioned, reaction temperature is a very important physical quantity which may affect methane output and overall reaction rate. Each organism in the fermentation mixture has an optimum temperature where growth is maximal, and upper and lower temperature bounds beyond which the population cannot survive (2).

Methanogenic bacteria may be classified according to activity in a temperature range into two groups, the mesophilic group and the thermophilic group. The mesophilic bacteria are active between 45 C and 65 C. When digesting sewage sludge, mesophilic bacteria show optimum activity between 35 C and 40 C, while thermophilic bacteria show optimum activity between 52 C and 60 C (8,10,13).

Based upon laboratory studies involving the anaerobic digestion of sewage sludge, different conclusions regarding the rates of gas production of mesophilic and thermophilic bacteria have been formulated. Some authors have observed a higher rate of gas production in the thermophilic range as compared to the rate of gas production in the mesophilic range, and vice versa. For example, Pfeffer (12) has observed a higher rate of gas production for the digestion of municipal refuse in mixed culture in the thermophilic range.

Contrary to the results obtained by Pfeffer, Malina (10) has observed that a higher rate of gas production for the digestion of sewage sludge in mixed culture results in the mesophilic range. For example, at a loading rate of 4.81 gVm/liter/day and retention time of 12 days, the rate of gas production varied from 1.05 l/gVm destroyed/day at 32.5 C to 0.76 l/gVm/day at 42.5 to 0.91 l/gVm/day at 52.5 C. Similar results were obtained for other conditions of loading and retention time, showing that the maximum gas production per unit of volatile matter destroyed occurred at 32.5 C, the minimum production at 42.5 C, and an intermediate value at 52.5 C. Pohland and Bloodgood (13) have also observed the maximum methane gas production in the mesophilic range, at 36 C, when digesting sewage sludge anaerobically.

It is evident that there is a controversy among authors when considering maximum gas production as a function of temperature in the mesophilic and thermophilic ranges. Since the rate of gas production plays a very important role in the economic analysis of the process, the effect of temperature on the anaerobic digestion process opens a very interesting and attractive research area.

It should be noted that any increase in gas production by operating the process in thermophilic range should offset the amount of energy required to maintain the reactor temperature at the higher temperature. For best operating performance in the thermophilic range, the amount of gas required to maintain the reactor temperature at the higher temperature subtracted from the total gas produced should still be higher than the rate of gas production in the mesophilic range. This is necessary to make the operation of the process in the thermophilic range economically attractive.

The purpose of this study is to study the effects of temperature on the anaerobic digestion of crop material to methane. Is the rate of methane production greater in the mesophilic or thermophilic region, and why? Can any increase in methane production offset the additional heat requirements for the reactor? What happens in the transitional region between mesophilic and thermophilic ranges? These questions will be answered in this paper. In addition, kinetic models will be presented at the various operating temperatures and an economic analysis will be presented as a function of operating temperature and reactor size.

PROCEDURE

The anaerobic digesters used in this study were four 7.0 liter Plexiglass stirred-tank reactors housed in a constant temperature cabinet. A schematic diagram of a digester is shown in Fig.1. These reactors were specifically designed to ensure that the process operated anaerobically by preventing the leakage of oxygen (air) into the reactor systems from the outside. Each digester had an inlet for the daily introduction and removal of feed slurry, and was equipped with a pitched-blade paddle agitator with variable speed drive. The product gas exited from the top of the reactor and was collected and measured daily in a water displacement gas meter, as described by Clausen and Gaddy (3). Hammer-milled cornstalks were chosen as a representative feed material for this study because of their abundance and high rate of growth. Sewage sludge from an anaerobic primary digester was used as the source of bacteria. In order to obtain kinetic data as a function of temperature the reaction temperature was varied from 35 to 60 C at five degree intervals, and the feed concentration ranged from 2 to 10 percent by weight solids. Retention time in the mixed flow reactor remained at 22 days.

Gas composition and pH were checked daily and periodic measurements of ammonia nitrogen and phosphate concentrations were also taken. Gas composition was measured using gas-solid chromatography (1/8" x 10' Porapak QS column) and ammonia nitrogen and phosphate concentrations were measured using a spectrophotometric technique. Adequate operation time at a given condition was allowed in order to reach steady state so that accurate kinetic data could be taken.

RESULTS

Steady-state values of gas production at various feed concentrations as a function of reaction temperature are shown in Fig. 2. The maximum rates of gas production in the mesophilic temperature range for the feed concentrations ranging from 2.0 to 10.0 percent occurred at 40 C, while in the thermophilic temperature range the maxima occurred at 55 C. At 40 C, average production rates of 2.009 liters/day to 6.896 liters/day of gas were observed when varying the feed concentration from 2.0 to 10.0 percent. At 55 C, average production rates of 0.683 liters/day to 2.825 liters/day of gas were observed when varying the feed concentration from 2.0 to 10.0 percent. Minimum production rates were shown to occur in the temperature range from 45 to 50 C for all feed concentrations.

Several interesting statements may be made in reference to the behavior shown in Fig. 2. As expected, gas production increased with increasing feed concentration except at the transitional temperature, 45 C. The increase in gas production, however, was not not directly proportional to feed concentration, but tended to increase only slightly at higher feed concentrations.

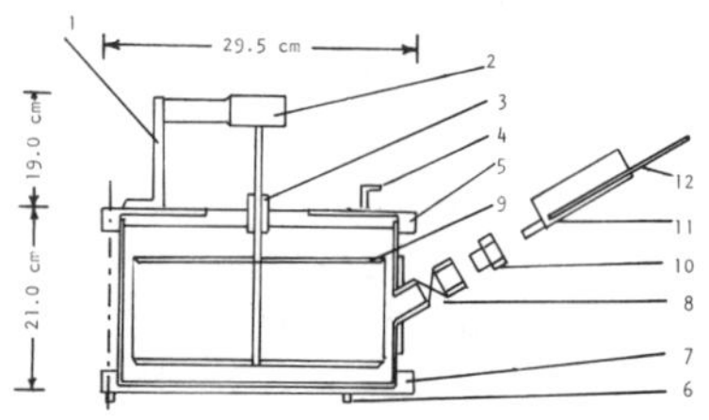

1. L-support
2. A.C. motor
3. Grease seal
4. Outlet for the gas produced
5. Top aluminum flange
6. Leg-support for the digester
7. Bottom aluminum flange
8. Feed inlet (and sludge outlet)
9. Agitator
10. Aluminum adapter
11. Plexiglass feeding cylinder
12. Feeding metal rod

Figure 1. Schematic Diagram of the Digester

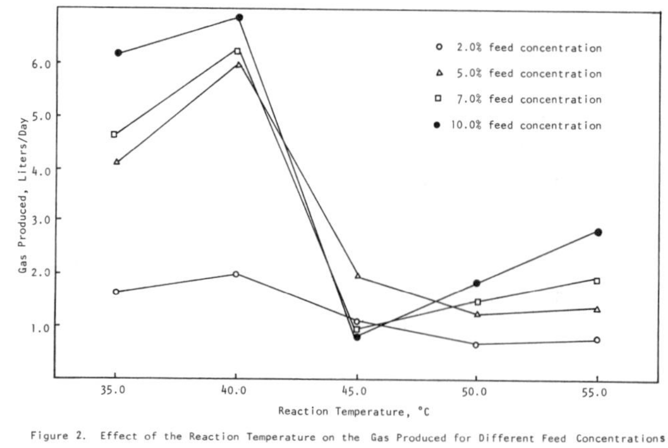

Figure 2. Effect of the Reaction Temperature on the Gas Produced for Different Feed Concentrations

The rate of gas production was lowest at 45 C, which corresponds to the transitional region between optimum mesophilic and thermophilic operation. Neither mesophilic or thermophilic bacteria are dominant in this range, so that overall gas production rates are decreased.

The third major observation is that the rate of gas production in the mesophilic range (40 C) is greater than the maximum value studied in the thermophilic range (55 C). Three possibilities exist for explaining this behavior: 1. recovery after operation at 45 C was never quite complete thus lowering the maximum rate at 55 C; 2. the optimum temperature in the thermophilic range was not reached and lies above 55 C; or 3. operation in the mesophilic range gives superior methane yields in comparison to operating in the thermophilic range.

The correct conclusion remains uncertain when considering the existing data only. Operation at temperatures above 55 C was not possible due to equipment limitations. However, it is significant to note that reaching operating temperatures above 45 C presents very real problems since process instabilities are apparent in the transitional region. These problems may cause thermophilic operation to be difficult or undesirable unless another source of methane bacteria (other than sewage sludge) is used.

Average methane concentrations in the product gas for the conditions shown in Fig. 2 are presented in Table I. Although general conclusions about this data are difficult to formulate, the lower values of methane composition are definitely seen at 45 to 50 C, and methane concentration does appear to decrease with increasing feed concentration.

It is interesting to compare these results with a batch digestion study made by Klass, et al (8). He observed maximum rates of gas production at 40 C in the mesophilic range and at 60 C in the thermophilic range while studying the anaerobic digestion of grass in a mixed-cultured batch operation. Also, higher rates of gas production were observed at 40 C in comparison to the rates of gas production at 60 C. The study concluded that mesophilic conditions are preferred to thermophilic conditions for the production of methane gas by the anaerobic digestion process.

Table I. Steady-state Methane Compositions in the Gas Produced as a Function of Feed Concentration and Reaction Temperature

	Methane Percentage in Gas Produced				
Reaction Temperature	35 C	40 C	45 C	50 C	55 C
Feed Concentration 2.0% by weight	58.34	58.32	58.44	54.64	61.35
Feed Concentration 5.0% by weight	50.71	53.17	49.16	50.37	54.14
Feed Concentration 7.0% by weight	47.43	52.32	49.07	47.12	51.05
Feed Concentration 10.0% by weight	47.48	46.88	45.64	46.43	49.53

KINETIC STUDIES

Since a complex cellulosic substrate (cornstalks) was used in the study, a method was needed to follow substrate utilization. Cornstalks were found to contain 45 percent carbon, and carbon concentration was used as the limiting substrate to be followed in both the inlet and outlet streams to the reactor. Reaction rate was calculated by the equation

$$-r_C = \frac{C_{so} - C_s}{\theta} \qquad (1)$$

where $-r_C$ = reaction rate,

C_{so} = inlet carbon concentration

C_s = outlet carbon concentration,

and θ = retention time.

A plot of reaction rate as a function of carbon concentration in the digester is shown in Fig. 3. As with classical kinetics, the reaction rate was found to increase with increasing carbon concentration in the digester at all temperatures of the study except 45 C. At 45 C, the reaction rate for feed concentrations of 7.0 and 10.0 percent were observed to be lower than that for the 5.0 percent feed concentration. This is probably due to the instability of the process at the transitional temperature of 45 C. The maximum reaction rates for a fixed feed concentration were observed at 40 C, with a maximum reaction rate of 0.63 grams of carbon/liter/day observed for a 10.0 percent feed concentration at 40 C.

Classically, the fermentation kinetics of reactions such as the anaerobic digestion reaction can be represented by

$$-r_c = \frac{k_1 \cdot C_m \cdot C_c}{k_2 + C_c} \qquad (2)$$

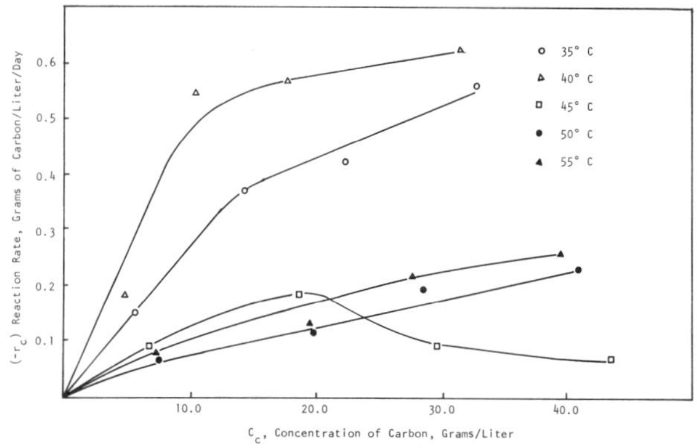

Figure 3. Effect of the Carbon Concentration on the Reaction Rate at Different Temperatures

where C_c = concentration of the limiting substrate;
C_m = concentration of organisms in the bacterial culture;
k_1 = reaction rate constant; and,
k_2 = reaction rate constant.

If the assumption is made that C_m, the concentration of organisms, is nearly constant, Equation (2) may be expressed in the form

$$-r_c = \frac{k_1' \cdot C_c}{K_2 + C_c} \tag{3}$$

where $k_1' = k_1 \cdot C_m$.

This type of equation is preferred in this analysis due to the slurry nature of the reaction mass, thus making cell analysis difficult and unreliable.

A plot of $(-1/r_c)$ as a function of $(1/C_c)$ is shown in Fig. 4. As shown, at a fixed temperature a plot of $(-1/r_c)$ vs. $(1/C_c)$ shows straight line behavior and, hence, Equation (3) is satisfactory for this reaction system.

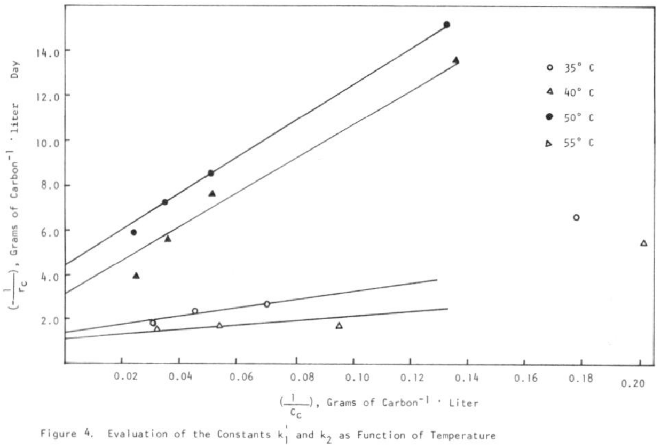

Figure 4. Evaluation of the Constants k_1' and k_2 as Function of Temperature

The date at 45 C did not show this trend and were not plotted. The model shown in Equation (3) should not be expected to fit this transitional data.

The reaction rate constants k_1' and k_2 in Equation (3) at the different temperatures of study were evaluated by using a least squares straight line computer programming technique. The constants k_1' and k_2 with their standard deviations are presented in Table II. Since a slight deviation from the kinetic model is shown at a 2 percent feed concentration, the constants were also evaluated when excluding this data. Table III presents the constants k_1' and k_2 and standard deviations evaluated by excluding the data for the 2.0 percent concentration. From a comparison of the standard deviations given in both of these tables, it is evident that it is more reliable to use the constants k_1' and k_2 given in Table III for feed concentrations above 2.0 percent by weight. Also, the use of low feed concentrations such as 2.0 percent by weight do not have any practical applications because of their considerable lower gas production rates and higher fixed and operating costs for a digester, while considering the same production capacity.

PROCESS DEVELOPMENT AND ECONOMICS

Utilizing the data previously presented, a preliminary process design may be constructed for manufacturing methane from agricultural residues. For the purposes of this study, a plant size was chosen utilizing 200 metric tons of cornstalks per day. Fig. 5 presents the necessary processing steps and equipment for this process.

Table II. Reaction Rate Constants k_1' and k_2 as a Function of Reaction Temperature

Reaction Temperature, C	k_1', $\frac{grams}{liter \cdot day}$	k_2, $\frac{grams}{liter \cdot day}$	Standard Deviation
35.0	1.40	45.6	0.2739
40.0	2.53	60.2	0.6963
45.0	--	--	--
50.0	0.25	20.8	0.2411
55.0	0.40	33.9	0.8039

Table III. Reaction Rate Constants k_1' and k_2 as a Function of Reaction Temperature*

Reaction Temperature, C	k_1', $\frac{grams}{liter \cdot day}$	k_2, $\frac{grams}{liter}$	Standard Deviation
35.0	0.81	17.4	0.2162
40.0	0.66	2.3	0.0572
45.0	--	--	--
50.0	0.28	27.6	0.2135
55.0	3.15	455.4	0.0705

*Reaction rate data for 2.0% feed concentration are not considered for the evaluation of reaction rate constants k_1' and k_2.

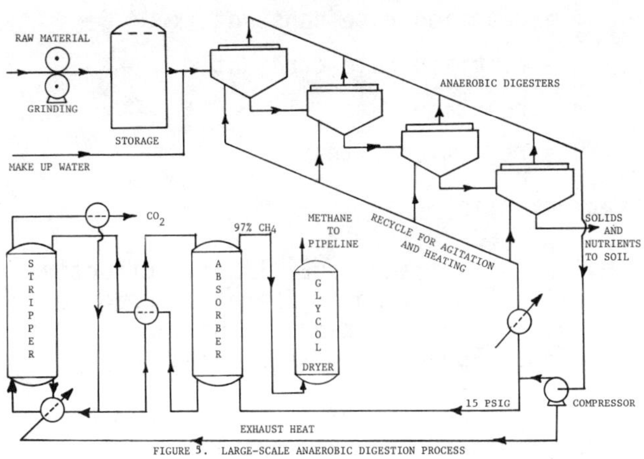

FIGURE 5. LARGE-SCALE ANAEROBIC DIGESTION PROCESS

The crop matter is passed through a shredder before entering the reactors. Storage silos provide one day of feed retention. Ground biomass is mixed with water before entering the reactors. The reactors are floating-head steel-insulated tanks, with heating and agitation provided by gas recirculation.

The product gas stream is compressed to 15 psig. Carbon dioxide and hydrogen sulfide are scrubbed from the methane with monoethanolamine solution. A glycol scrubber dries the gas to produce pipeline quality methane.

By making use of the kinetic analyses and the process design presented in Fig. 5, a preliminary economic analysis may be made at each operating temperature. A representative economic analysis at a temperature of 40 C is presented in Table IV using a feed concentration of 7 percent by weight. Capital cost estimates are based upon data presented by Guthrie (6) and Peters and Timmerhaus (11). All estimates were updated to the First Quarter, 1980, by using the Marshall and Stevens index.

Table IV

Process Economics for Converting Crop Material To Methane

Temperature: 40°C

Capital Investment	
Digesters	$ 1,384,000
Milling (sample preparation)	11,000
Compressors	332,000
Pumping and piping	129,000
MEA stripper and absorber	58,000
Heat exchangers	167,000
Contingency (30%)	624,000
Total	$ 2,705,000
Operating costs (in dollars/year)	
Raw material costs (collection and transportation costs)	$ 803,000
Utilities	37,000
Labor	136,000
Maintenance	136,000
Depreciation, Taxes, and Insurance	271,000
Total	$ 1,383,000/yr
Methane Cost (Break-Even)	$ 0.10/M liters ($2.95/M ft^3)

As shown in Table IV, the total capital investment for a plant using 200 metric tons of cornstalks per day is $2.7 million including a 30 percent contingency. Over 50 percent of this investment is for the digesters, with the compressor, milling costs, pumping costs, MEA stripper, and heat exchangers accounting for the balance.

Yearly operating costs amount to $1.383 million, of which $803,000 is for the collection and transportation of crop waste to the plant site. Harvesting and transportation costs are based on a study that shows biomass can be produced and harvested for $11 per metric ton, (1). Maintenance costs and labor costs were each calculated as 5 percent of the investment, and depreciation, taxes and insurance were calculated as 10 percent of the investment. Operating costs for utilities amounted to $37,000 per year.

Based upon these operating costs, the break-even point for the methane produced is $0.10 per thousand liters ($2.95 per thousand cubic feet).

Similar studies were completed at a 7 percent feed concentration for temperatures of 35, 45, 50, and 55 C using the same plant size. The results of these studies are shown in Fig. 6. The minimum methane cost was at 40 C ($0.10 per thousand liters of methane) and the maximum was at the transitional temperature of 45 C ($0.77 per thousand liters of methane). Major contributing factors accounting for these results are the methane yields found in the study and the larger heat losses from the digesters as the temperature is increased. The percentages of the methane reproduced in the plant that is required to offset reactor heat losses were 4, 3, 22, 17, and 13 percent at the temperatures of 35, 40, 45, 50, and 55 C, respectively.

Methane break-even price also decreases with increasing plant size. This trend is shown in Fig. 7, where break-even methane price is shown to approach about $0.07 per thousand liters ($2.00 per thousand cubic feet) as the plant size is increased significantly to a feed rate of 10000 metric tons of crop waste per day.

CONCLUSIONS AND SIGNIFICANCE

Reaction temperature has been shown to significantly affect the kinetics of methane production as well as the price of methane production. An optimization study involving reaction temperature, however, is not in it-

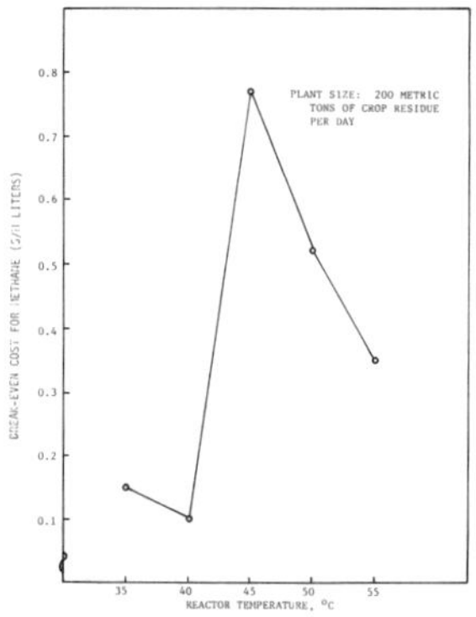

FIGURE 6. BREAK-EVEN COST AS A FUNCTION OF REACTION TEMPERATURE

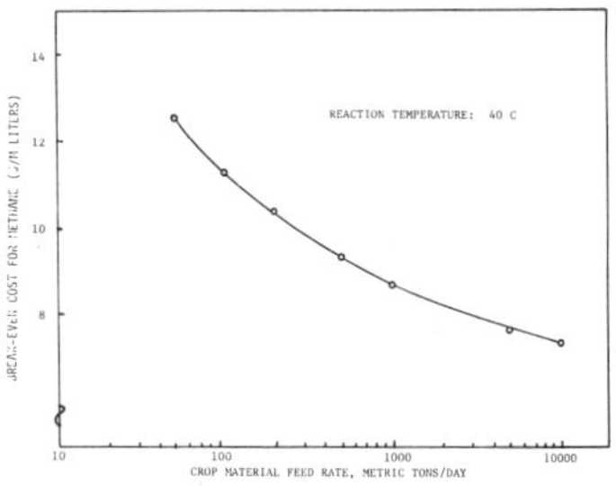

FIGURE 7. BREAK-EVEN COST AS A FUNCTION OF PLANT SIZE

self enough to make methane production for biomass economically attractive. Large plant sizes and optimization of other process variables are still necessary to make the operation of a methane plant feasible in today's energy market.

As previously mentioned, a definite conclusion in regard to thermophilic operation is not possible using the results of this study. It is accurate to say, however, that reaching the optimum temperature in the thermophilic range using a culture of domestic sewage sludge presents many operating difficulties due to process instabilities in the transition region.

Methane production from biomass should, with continued research, become economically attractive provided the research is aimed at decreasing capital and operating expenses by increasing reaction rates.

NOTATION

C_c = concentration of limiting substrate
C_m = concentration of organisms in the bacterial culture
C_s = outlet carbon concentration
C_{so} = inlet carbon concentration
k_1 = reaction rate constant
k_1' = reaction rate constant, $k_1 \quad C_m$
k_2 = reaction rate constant
$-r_c$ = reaction rate
θ = retention time

LITERATURE CITED

1. Alich, J. A., and R. E. Inman, "Effective Utilization of Solar Energy to Produce Clean Fuel," prepared for National Science Foundation, RANN, Final Report (1974).
2. Bailey, J. A., and D. F. Ollis, Biochemical Engineering Fundamentals, 1st ed., McGraw-Hill, New York (1977).
3. Clausen, E. C., and J. L. Gaddy, "Separate-Stage Fermentation of Biomass to Methane," AIChE Symposium Series, 181, Vol. 74, (1978).
4. Clausen, E. C., O. C. Sitton, and J. L. Gaddy, "Biological Production of Methane from Energy Crops," Biotechnology and Bioengineering, 21 (7), 1209 (1979).
5. Fullen, W. J., "Anaerobic Digestion of Packing Plant Wastes," Sewage and Industrial Wastes, 25 (5), 576 (1953).
6. Guthrie, K. M., "Capital Cost Estimating," Chemical Engineering 76 (6), 114 (1969).
7. Klass, D. L., "A Perpetual Methane Economy," Gas Technology, (1974).
8. Klass, D. L., S. Ghosh, and J. R. Conrad, "The Conversion of Grass to Fuel Gas for Captive Use," Symposium Papers on Clean Fuels from Biomass, Sewage, Urban Refuse, and Agricultural Wastes, Institute of Gas Technology, 223, January 1976.
9. Loehr, R. C., and R. W. Agnew, "Cattle Wastes - Pollution and Potential Treatment," ASCE Proc. J. San. Eng. Div., 93 (SA4), 55 (1967).
10. Malina, J. F., Jr., "The Effect of Temperature on High-Rate Digestion of Activated Sludge," Proceedings of the 16th Industrial Waste Conference, Purdue University, Lafayette, IN (1961).
11. Peters, M. S., and K. D. Timmerhaus, Plant Design and Economics for Chemical Engineers, 2nd ed., McGraw-Hill, New York (1968).
12. Pfeffer, J. T., "Reclamation of Energy from Organic Waste," prepared for Environmental Protection Agency, NTIS Bulletin No. PB-231, 176 (March, 1974).
13. Pohland, F. G., and D. E. Bloodgood, "Laboratory STudies on Mesophilic and Thermophilic Anaerobic Sludge Digestion" Journal of Waste Pollution Control Fed., 35 (1), 11 (1963).
14. Ricci, L. J., "Garbage Routes to Methane,"

Chemical Engineering 81 (11), 58 (1974).
15. Schroepfer, G. J., W. J. Fullen, A. S. Johnson, N. R. Ziemke, and J. J. Anderson, "The Anaerobic Contact Process as Applied to Packinghouse Wastes," Sewage and Industrial Wastes, 27 (4), 460 (1955).

COMPUTER SIMULATION OF SOLAR COOLING

Numerical simulation of a solar operated water-lithium bromide absorption system is presented and tested. A mathematical model is prepared for each component of the proposed system and for the solar radiation, ambient conditions, and the properties of the working fluids as well. The system is found to be simulated by a set of twenty four simultaneous equations. An easy-to-follow procedure is outlined to enable the user to get a transient solution of these equations under specified conditions. Newton-Raphson technique is used to handle the nonlinearity of the governing equations when applicable. Numerical performance tests are carried out for a sample summer day and a sample winter day from 8:00 AM to midnight. The results showed that the technique is both reliable and easy to follow.

M. M. ELSAYED
I. S. TAHA
and
M. A. DARWISH
Mechanical Engineering Department
King Abdulaziz University
P.O. Box 9027
Jeddah, Saudi Arabia

INTRODUCTION

Simulation is a powerful tool in the analysis of an engineering process. In simulation, mathematical descriptions of the individual components are integrated into a dynamical system to determine the process in response to specified input conditions. Simulation can be used to inform about the component interaction, sensitivity to changes of design parameters, system behaviour, ...etc. It can be used as a substitute for performing experiments to obtain information about the process variable parameters. It is used extensively as a design tool and several simulation programs have been developed for such purpose.

Simulation is used extensively in solar operated hot water or heating systems. Some of the available simulation programs were extended to include solar operated cooling systems. A survey of the existing solar simulation programs was reported by Versteegen et al (1). Some of the well known published programs-e.g. DOE-1 (2), TRANSYS (3), SOLSEM (4), SINSHAC (5),...etc-are very sophisticated programs performing detailed hour by hour analysis. Those programs are very expensive to use and need extensive input data. Each of these programs has certain limitations and proved to be useful in specific applications.

The need for simple simulation programs is obvious for practicing and design engineers since very limited number of these simple programs are available. Also, most of the available programs are usually reporting some of the equations used and the results obtained by running these programs on computer for specified inlet conditions. In most cases, no procedures (or illustration)-to-follow,to establish similar programs are given.

The aim of present work is to illustrate a simple and straight forward technique to develop a simulation analyzing a solar-operated H_2O/Li Br absorption cooling system. The simulation performs heat and mass balances over different sections of the system. The used method is certainly applicable to establish simulation programs for other engineering processes, especially chemical engineering processes, where heat and mass transfer are involved in different components of the system.

MODELLING OF DIFFERENT COMPONENTS

Consider the water-lithium bromide refrigeration system illustrated

0065-8812-81-6039-0210-$2.00

in Figure 1. Solar radiation is used

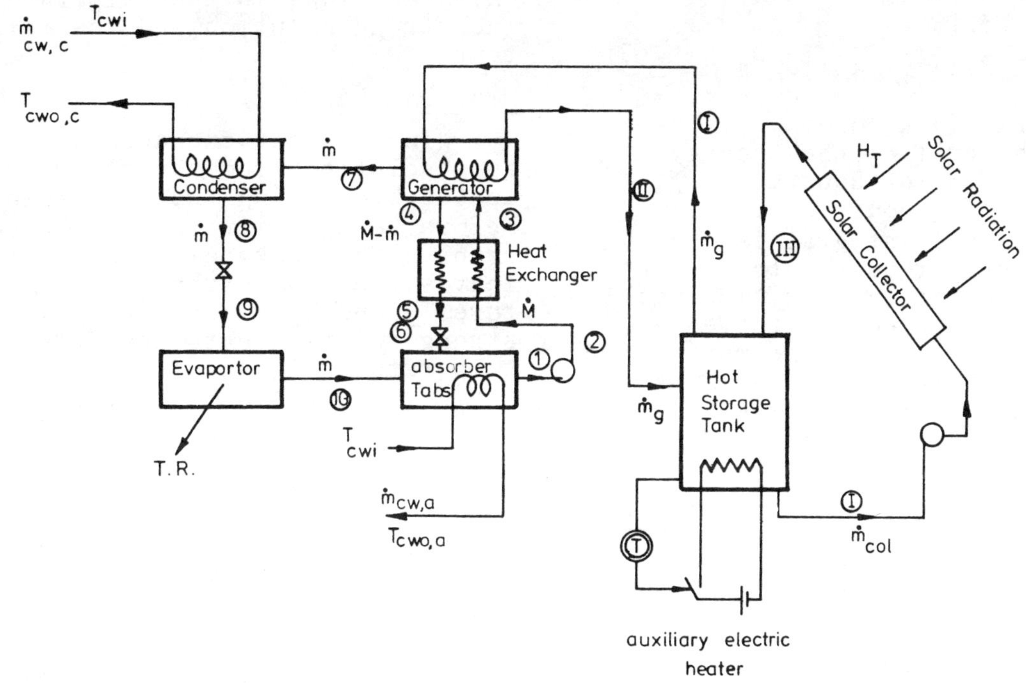

Figure 1 Flow diagram of the solar-operated water/Li Br absorption cooling machine

to heat a fluid flowing in a flat plate collector. The heated fluid from the collector is then stored in a hot storage tank for reuse as a heating medium for the generator of the absorption machine as shown in the figure. The solar cooling system consists of the following components: collector, hot storage tank, collector circulating pump, auxiliary heater for the hot storage tank, circulating pump of the heating medium to the generator, absorber, water-lithium bromide circulating pump, heat exchanger, generator, strong mixture expansion valve, condenser, refrigerant expansion valve, and evaporator. Modelling of each component is considered separately below.

Collector

The considered collector is a flat plate type having an area A_c . The glass covers have a total transmissivity τ . The absorber of the collector has an absorptivity α . The absorbed heat is transferred partly to the fluid which enters the collector at temperature T_I and leaves at temperature T_{III}. The fluid is circulated through the collector at a rate of $\dot{m}_{col}$. The net heat rate absorbed by the heat removing fluid Q_{col} is related to the previously mentioned parameters and the surroundings' temperature T_a by the following equations

$$Q_{col} = \dot{m}_{col} \, C_{p,col} \, (T_{III} - T_1) \quad (1)$$

$$Q_{col} = A_{col} \, F_R(H_T \tau \alpha - U_L(T_I - T_{am}))(2)$$

where F_R is the collector's heat removal factor, U_L is the overall heat loss coefficient from collector plate to the surroundings, $C_{p,col}$ is the specific heat of the collector fluid, and H_T is the total solar radiation flux incident on the glass cover surface.

Hot Storage Tank

The instantaneous temperature of the fluid in the hot storage tank T_I is governed by Q_{col} , the rate of heat supplied to the generator Q_g, and the heat capacity of the fluid in the tank $m_{HS}\ C_{v,col}$ according to the relation

$$m_{HS}\ C_{v,col} \frac{d\ T_I}{d\ t} = Q_{col} - Q_g \qquad (3)$$

The size of the tank, or m_{HS}, is estimated by integrating the energy stored, the R.H.S. of Equation (3), over the operating hours during sunshine. Then, this integral is equated to the energy required to run the machine over an interval of time with no solar input. Moreover, there should be limitations on the maximum and minimum temperatures in the tank. The maximum temperature $T_{I,max}$ is chosen below the fluid boiling temperature. A minimum temperature of the fluid in the storage tank, $T_{I,min}$ is set to insure reasonable performance of the cooling machine.

Absorber

The absorber is assumed to have an overall heat transfer coefficient-area product $(UA)_a$ and cooling water flow rate $\dot{m}_{cw,a}$. The cooling water enters and leaves the absorber at temperatures T_{cwi} and $T_{cwo,a}$ respectively. The water-Lithiumbromide strong solution of concentration x_4 enters the absorber at state (6) with mass flow rate $(\dot{M}-\dot{m}_7)$, the refrigerant water enters at state (7) with mass flow rate $\dot{m}_7$, and the weak solution of concentration x_1 leaves the absorber with mass flow rate $\dot{M}$. A mass and heat balances for the absorber give

$$\dot{m}_7 = (1 - \frac{x_1}{x_4})\ \dot{M} \qquad (4)$$

$$Q_a = \dot{m}_{cw,a}\ c_{p,w}(T_{cwo,a}-T_{cwi}) \qquad (5)$$

$$Q_a = \dot{M}\ (h_{10}-h_1) + \frac{x_1}{x_4}\ \dot{M}(h_6-h_{10}) \qquad (6)$$

$$Q_a = (UA)_a\ \ \theta_{m,a} \qquad (7)$$

where

$$\theta_{m,a} = \frac{(T_{abs} - T_{cwi}) - (T_1 - T_{cwo,a})}{\ell n(\frac{T_{abs} - T_{cwi}}{T_1 - T_{cwo,a}})} \qquad (8)$$

and

$$T_{abs} \cong T_6 \qquad (9)$$

Also, $\dot{m}_{cw,a}$ is mass flow rate of the cooling water circulated in the absorber, h_n is the enthalpy at state (n), and T_{abs} is the temperature of the strong solution and the refrigerant.

Heat Exchanger

In the heat exchanger, the strong solution from the generator is cooled by heating the weak solution before being delivered to the generator. The weak solution enters and leaves the heat exchanger at temperatures T_2 and T_3 respectively while the strong solution enters and leaves at temperatures T_4 and T_5 respectively. The governing equations for the heat exchanger are

$$Q_{ex} = \dot{M}\ C_{p,2}\ (T_3 - T_2) \qquad (10)$$

$$Q_{ex} = \frac{x_1}{x_4}\ \dot{M}\ C_{p,4}\ (T_4 - T_5) \qquad (11)$$

$$Q_{ex} = (UA)_{ex}\ \ \theta_{m,ex} \qquad (12)$$

$$\theta_{m,ex} = \frac{(T_4 - T_3) - (T_5 - T_2)}{\ell n\ (\frac{T_4 - T_3}{T_5 - T_2})} \qquad (13)$$

where Q_{ex} is the heat transferred from the strong solution to the weak solution, and $(UA)_{ex}$ is the product of overall heat transfer coefficient and the area for the heat exchanger.

Generator

The generator of the absorption machine is fired by hot water supplied from the hot storage tank at mass flow rate of $\dot{m}_g$. The hot water enters and leaves the generator at temperatures T_I and T_{II} respectively. The governing equations for the generator are

$$T_7 = T_4 \qquad (14)$$

$$Q_g = \dot{m}_g\ C_{p,col}(T_I - T_{II}) \qquad (15)$$

$$Q_g = \dot{M}\ (h_7 - h_3) + \frac{x_1}{x_4}\ \dot{M}(h_4 - h_7) \qquad (16)$$

$$Q_g = (UA)_g\ \ \theta_{m,g} \qquad (17)$$

$$\theta_{m,g} = \frac{(T_I - T_3) - (T_{II} - T_4)}{\ln\left(\frac{T_I - T_3}{T_{II} - T_4}\right)} \quad (18)$$

where $(UA)_g$ is the product of the overall heat transfer coefficient and the area for the generator.

Condenser

The condenser of the absorption machine is selected to be water cooled. The mass flow rate of the cooling water is controlled in a way to keep 1 ^oC subcooling at the refrigerant exit condition. Thus, the overall heat transfer coefficient between the cooling water and the refrigerant is affected by the change in mass flow rate. The governing equations for the condenser can be summarized as follows

$$Q_c = \dot{m}_7 (h_7 - h_8) \quad (19)$$

$$Q_c = \dot{m}_{cw,c} C_{p,w} (T_{cwo,c} - T_{cwi}) \quad (20)$$

$$Q_c = (UA)_c \quad \theta_{m,c} \quad (21)$$

$$\theta_{m,c} = \frac{(T_8 - T_{cwo,c}) - (T_7 - T_{cwi})}{\ln\left(\frac{T_8 - T_{cwo,c}}{T_7 - T_{cwi}}\right)} \quad (22)$$

$$(UA)_c = 21990/(6.046 + 18.93 \dot{m}_{ref}^{1/3} + 5.659 \dot{m}_{cw,c}^{-0.796}) \quad (23)$$

$$T_8 = T_{sat} - 1 \quad (24)$$

where T_{sat} is the saturation temperature of the refrigerant in the condenser. Equation (23) was derived from an equation given by Reference (8).

Evaporator

It is assumed that the refrigerant leaves the evaporator at dry saturation condition. No simulation for the load will be considered. The cooling effect Q_e is to be calculated from

$$Q_e = \dot{m}_7 (h_{10} - h_9) \quad (25)$$

Strong Mixture and Refrigerant Expansion Valves

Both expansion valves are assumed to be well insulated. The energy equation yields

$$h_6 = h_5 \quad (26)$$

$$h_9 = h_8 \quad (27)$$

SUPPORTING EQUATIONS

In the previous section, modelling of the various components had been developed. In order to carry out a performance test for the system, supporting equations prescribing the properties of the binary solution, the refrigerant, the solar radiation, and the ambient temperature have to be developed.

Properties of the Binary Solution

The equilibrium temperature (at a certain pressure) T^* and the specific heat of the binary solution C_p were correlated with the concentration x using the available data (9) via a least square technique. The resulting relations are

$$T^* = c_1 + c_2 x \quad (28)$$

$$C_p = b_1 + b_2 x \quad (29)$$

where c_1 and c_2 are functions of the pressure of the binary solution whereas b_1 and b_2 are constants. The enthalpy of subcooled states of the binary solution are obtained by multiplying the specific heat C_p by the temperature of the solution. In case of state (4), the temperature is the equilibrium temperature of the generator.

Properties of Refrigerant

Fixed refrigerant states in the condenser and the evaporator are assumed in this modeling. Thus, the enthalpies at states (8) and (10) are determined from thermodynamic tables and are used as input data. The enthalpy of the superheat state (7) is correlated with the temperature (at a certain pressure) using steam tables and a least square technique. The resulting equation is

$$h = a_1 + a_2 T$$

where both a_1 and a_2 depend on the pressure of the refrigerant.

Solar Irradiation

The instantaneous solar flux H_T is estimated for Jeddah location using the following equations, (6) and (7)

$$H_T = I_o \, Exp(\frac{-B}{\sin \beta}) \, (\cos \theta + C \, F_{ss}) \quad (31)$$

where

$$\cos \theta = \cos \beta \cos \gamma \sin s + \sin \beta \cos s \quad (32)$$

$$\beta = \sin^{-1}(\cos L \cos \delta \cos \omega + \sin L \sin \delta) \quad (33)$$

$$\gamma = |\Gamma + \phi| \quad (34)$$

$$\left.\begin{aligned} \Gamma &= \cos^{-1}\{\sec \beta (\cos \delta \cos \omega \sin L - \sin \delta \cos L)\} \\ &= 0 \text{ at solar noon for } L > \delta \\ &= \Pi \text{ at solar noon for } L < \delta \\ &= \text{undefined at solar for } L = \delta \end{aligned}\right\} \quad (35)$$

$$\delta = 23.45 \sin(360 \frac{284 + n}{365}) \quad (36)$$

$$\omega = 15 \, | t_s - 12 | \quad (37)$$

$$t_{ss} = \frac{1}{15} \cos^{-1}(-\tan \delta \tan L) + 12 \quad (38)$$

$$t_{sr} = 12 - \frac{1}{15} \cos^{-1}(-\tan \delta \tan L) \quad (39)$$

$$F_{ss} = \frac{1}{2} (1 + \cos s) \quad (40)$$

$$t_s = t_c + EOT + \frac{1}{15} (Lo - Lo_s) \quad (41)$$

All the symbols mentioned above are defined in the notation.

Air Temperature

The daily air temperature T_{am} is assumed to be a sin wave function of time t in the form

$$T_{am} = T_{av} + A' \sin \frac{\pi(t - t_{av})}{12} \quad (42)$$

where T_{av} is the arithmatic mean between the maximum temperature T_{max} and the minimum temperature T_{min} for a certain day. The amplitude of the wave A' is equal to $T_{am,max} - T_{am,min}$. The average time t_{av} in the arithmatic mean between the times corresponding to $T_{am,max}$ and $T_{am,min}$. These local standard civil times are assumed to be 3 p.m. (15) and 3 a.m. (3) respectively. The maximum and minimum temperatures for an average day for a certain month were estimated by judgement using data obtained from Jeddah weather station.

SYSTEM PERFORMANCE: NUMERICAL TEST

Each component of the cooling system shown in Figure 1 has been modeled by a set of equations. The performance of the system is determined by solving these equations simultaneously. However, fixed values have to be assigned to some variables in order to reduce the number of unknownsto the number of equations. The input data used for running performance tests are summarized in Tables 1 and 2.

Table 1 Input Parameters for Solar Radiation and Air Temperature Calculations

Day	Solar				Air Temperature	
	I_o W/m^2	B	C	EOT hr	$T_{am,max}$ ^oC	$T_{am,min}$ ^oC
June 11	1093.4	0.202	0.1300	0.00	40.2	21.0
December 16	1228.7	0.144	0.0590	0.100	32.9	14.8

Table 2 Input Parameters for Solar Cooling System

Device	Input Data
Collector	A_c = 300 m^2, $\tau\ \alpha$ = 0.88, F_R = 0.91, U_L = 2.7 $W/m^2\ {}^oK$, s = 21.5^o, $\dot{m}_c$ = 6 kg/s, Fluid : Water
Hot Storage Tank	Volume = 45 m^3, $T_{I,min}$ = 88 ^oC, $T_{I,max}$ = 105 ^oC
Generator	$\dot{m}_g$ = 19.4 kg/s, $(UA)_g$ = 12,220 W/^oK
Absorber	$\dot{M}$ = 1.5 kg/s, x_1 = 0.58 kg Li Br/kg, $(UA)_a$ = 6640 W/^oK, $\dot{m}_{cw,a}$ = 7.4 kg/s, T_{cwi} = 25 ^oC
Heat Exchanger	$(UA)_{ex}$ = 9,000 W/^oK
Condenser	T_{sat} = 38.78 ^oC, T_{cwi} = 25 ^oC, T_8 = 37.78 ^oC,
Evaporator	T_{10} = 4.44 ^oC, $(UA)_e$ = 4,750 W/^oK

The procedure of numerical solution is shown in Figure 2. The method uses successive loops iterations in a logic manner. Convergent solutions were obtained for all numerical test runs with small number of iterations.

Numerical results have been obtained for a sample summer day and a sample winter day for the period from 8 a.m. to midnight. It has not been intended to make a parametric study on the system, but just to check the constructed technique to simulate the system. The results of the two performance tests are shown in Figures 3 through 7.

The total solar input power and the cooling effect output are plotted versus time in Figure 3. The average cooling effects for the 16 hours of operation are 27.23 and 27.13 tons of refrigeration in the winter and summer days respectively. The variation in the cooling effect with time is small due to small variation in the storage tank temperature T_I as detected from Figure 4. This small variation in T_I is due to the large capacity of the hot storage tank (45 m^3). Figure 4 shows the outlet temperature from the collector T_{III}, the outlet heating water temperature from the generator T_{II}, the outlet weak solution temperature from the absorber T_1, and the outlet strong solution temperature from the generator T_4. The heat exchanger's UA, the minimum allowable temperature in the storage tank, and the outlet temperature from the cooling tower were so chosen that T_1 and T_4 are at a safe margin from equilibrium conditions. The maximum values of the strong solution concentration were 0.5997 and 0.5992 for the winter and summer days respectively.

The instantaneous mass flow rate of cooling water in the condenser of the cooling system is shown in Figure 5 together with the coefficient of performance of the absorption system (COP). The mass flow rate prediction is necessary for selection of pump and its control mechanism. The average values of the COP for sixteen hours of operation are 0.7104 and 0.7055 for the summer and winter days respectively.

External heating was needed sometimes in the morning and afternoon and during night to keep the temperature in the storage tank at 88 ^oC. The accumulated value of the external energy is shown versus time in the semilogarithmic plot of Figure 6. Also, the time-integrated incident solar irradiation and the time-integrated useful heat flow rate to water

Initial value of T_1 = minimum allowable temperature

- Assume T_1
- Assume T_3
- Assume x_4
- Calculate T_4, C_{p2} and h_4 → Eqs (28) , (29) and (30)
- C — Calculate Q_g , T_{II} , and $\theta_{m,g}$ → Eqs (16) , (15) and (18)
- Estimate x_4 by Newton-Raphson method → Eqs (16) and (17)
- Check the value of x_4 and replace till convergence
- A — Calculate T_4 , C_{p2} and h_4 → Eqs (28), (29) and (30)
- Calculate Q_g and T_{II} → Eqs (16) and (15)
- B — Calculate T_5 → Eqs (10) and (11)
- Calculate Q_{ex} and $\theta_{m,ex}$ → Eqs (10) and (13)
- Estimate T_3 by Newton-Raphson method → Eqs (10) and (12)
- D — Calculate T_5 → Eqs (10) and (11)
- Repeat till convergence of T_3 inside loop D
- Check T_3 and replace till convergence in its value in loop B
- Calculate Q_a and $T_{cwo,a}$ → Eqs (6) and (5)
- Calculate T_{abs} and $\theta_{m,a}$ → Eqs (9) and (8)
- E — Estimate T_1 by Newton-Raphson method → Eqs (6) and (7)
- Repeat till convergence of T_1 inside loop E
- Check T_1 and replace till convergence in its value in loop A
- Calculate T_I at the new time → Eqs (3) and (2)
- Repeat all the above procedure

Figure 2 Procedure of numerical tests

in the collectors are shown in Figure 6. At the end of the sixteen hours of operation, the values of the external energy to the collector useful energy ratio are 0.7302 and 0.8203 for the summer and winter days respectively. The hourly percentage of the generator energy that is obtained from solar energy , 100 × $(Q_g - Q_{ext})/Q_g$, is plotted at the middle of each hour and is given in Figure 7. The external energy can be decreased by decreasing the design minimum temperature of the storage tank. However, this will affect the performance of the system largely.

NOTATION

A	is area, m^2
a_1 , a_2	are constants dependent on the pressure of the refrigerant
B	is the atmospheric extinction coefficient, it depends on the day of the year.
b_1, b_2	are constants
C	is the uniform diffuse sky coefficient
C_p	is the specific heat at constant pressure, J/kg K
C_v	is the specific heat at constant volume, J/kg K
COP	is the coefficient of performance of the absorption machine
c_1, c_2	are constants dependent on the pressure of the binary solution
EOT	is the equation of time, hr
F_R	is the heat removal factor of the collector defined in Reference (7).

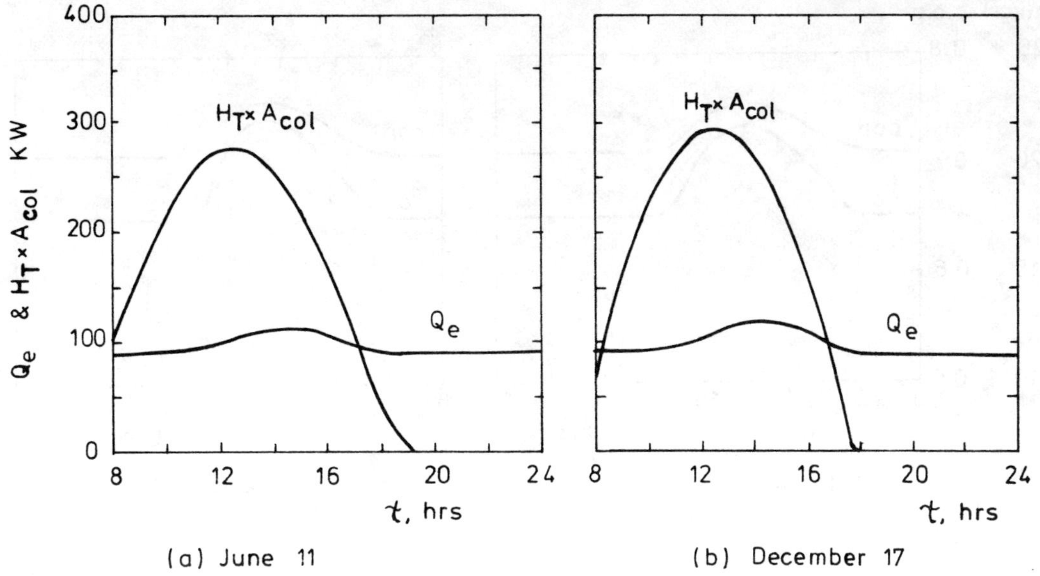

Figure 3 Instantaneous total solar irradiation and cooling effect

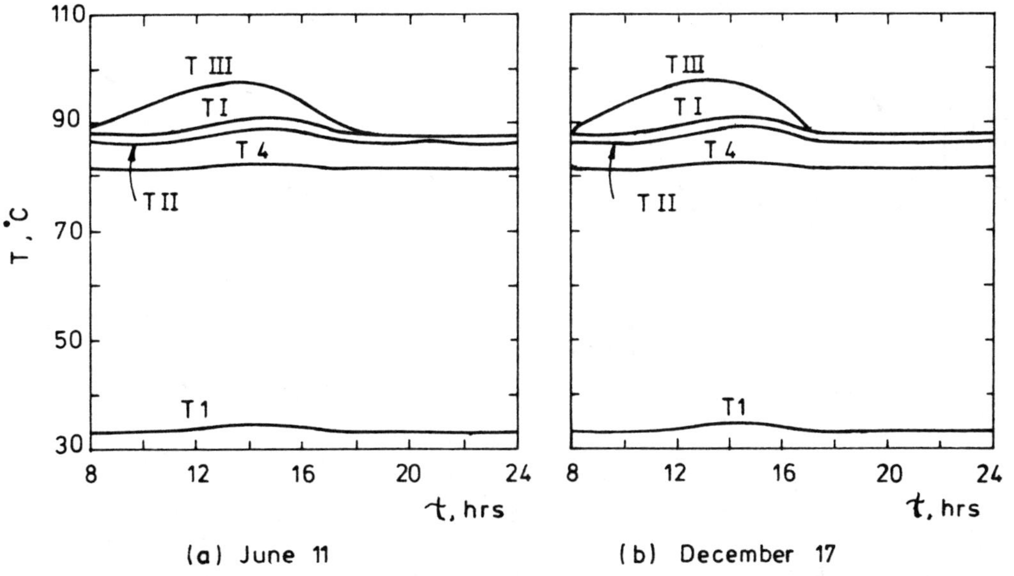

Figure 4 Instantaneous key temperatures for the collector, the generator, and the absorber

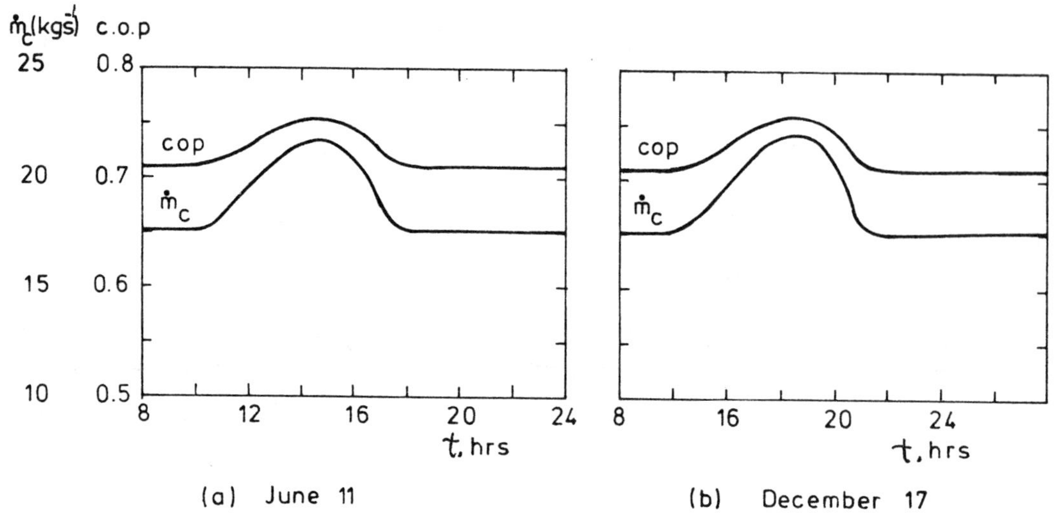

Figure 5 Instantaneous coefficient of performance and mass flow rate of cooling water in the condenser

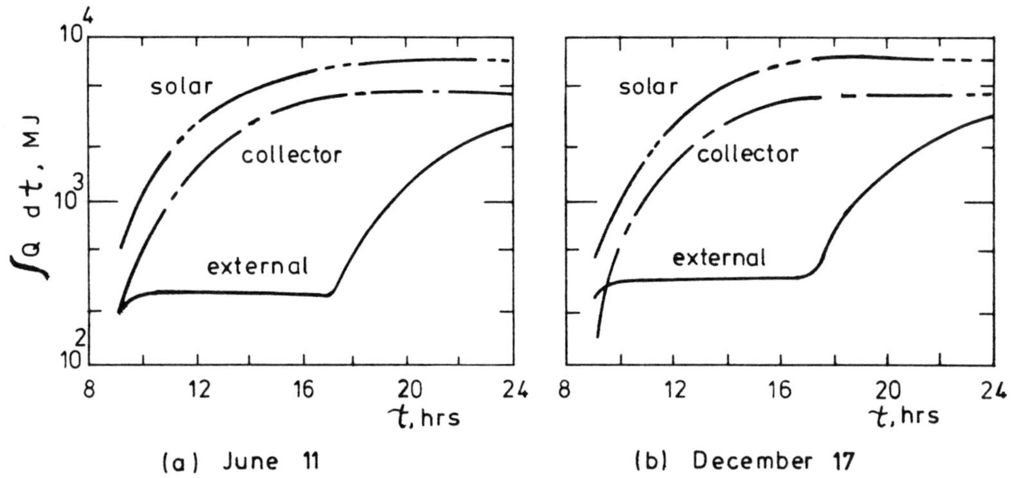

Figure 6 The time integral of solar power, of useful heat flow rate in collector, and of external heat flow rate to storage tank

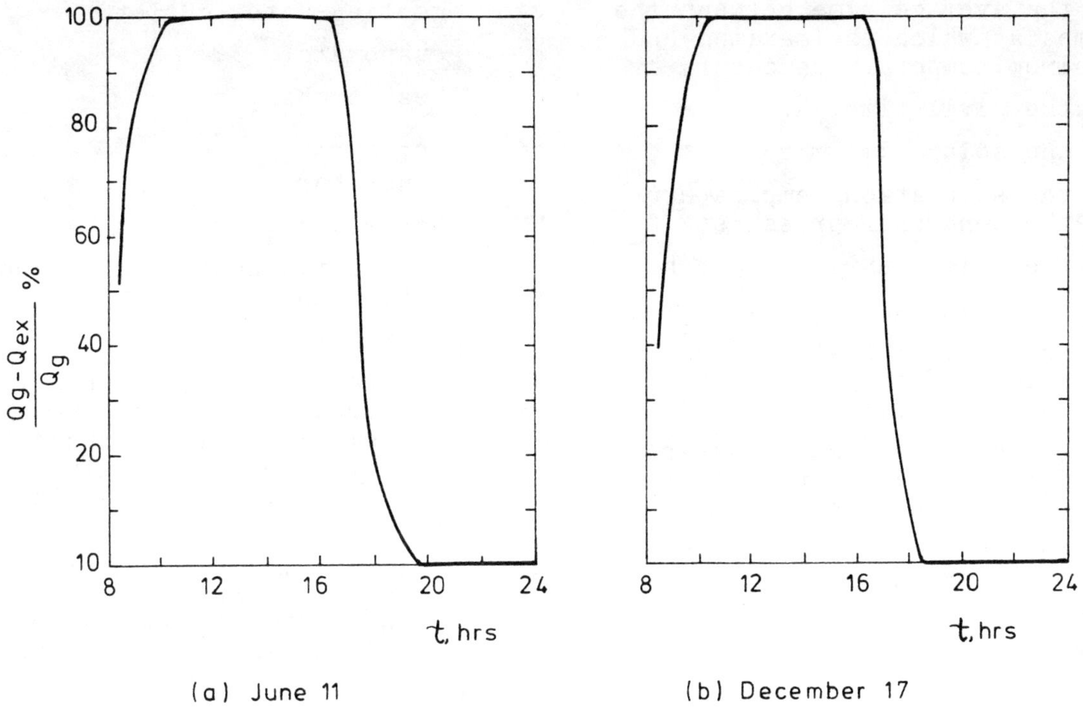

Figure 7 Instantaneous percentage of heat flow rate to generator which is obtained from solar energy

F_{ss}	is the surface-sky geometric factor
H_T	is the total (direct and diffuse) solar flux incident over a tilted surface, W/m^2
h	is the enthalpy, J/kg
I_o	is the solar constant, it depends on the day of the year, W/m^2
L	is the latitude angle (= 21.5 for Jeddah location)
Lo	is the east longitude angle of the location
Lo_s	is the east longitude angle of the location of the local standard time
$\dot{M}$	is the mass flow rate of the weak water-lithium bromide solution leaving the absorber, kg/s
$\dot{m}$	is the mass flow rate, kg/s
n	is the number of the day in the year
Q	is heat flow rate, W
s	is the collector surface tilt angle with the horizontal
T	is the temperature, K
T_{abs}	is the temperature of the strong solution and the refrigerant in the absorber, K
T_{av}	is the arithmatic average between the maximum and minimum temperatures for a certain day, ^oK
T_{max}	is the average maximum temperature for a certain month, K
T_{min}	is the average minimum temperature for a certain month, K
$T_{I,max}$	is the maximum temperature limit in the hot storage tank, K
$T_{I,min}$	is the minimum temperature limit in the hot storage tank, K
T^*	is the equilibrium temperature of the binary mixture at a certain pressure and certain concentration, K
	is the time, s

t_{av} is the average time between the times at which the maximum and minimum temperatures occur, hr

t_c is the civil time, hr

t_s is the solar time, hr

T_{sat} is the saturation temperature at the condenser pressure, K

t_{sr} is the solar sunrise time, hr

t_{ss} is the solar sunset time, hr

U_L is the overall heat loss coefficient of the collector defined in Reference (7), W/m^2 K

UA is the overall heat transfer coefficient area product of heat exchangers, W/ K

x is the concentration of the solution, kg Li Br/kg of solution

Greek Letters:

α is the absorptivity of the collectors' absorbers

β is the solar altitude angle

Γ is the solar azimuth angle measured from south (taken positive after solar noon and negative before solar noon)

γ is the wall-sun azimuth angle

δ is the solar declination angle in degrees

θ is the solar incidence angle

θ_m is the logarithmic mean temperature difference, ^oK

τ is the total transmissivity of the glass covers of each collector

ϕ is the collector orientation, it is positive east of south and negative west of south

ω is the solar hour angle, measured in degrees from solar noon

Subscripts

a absorber

am ambient

c condenser

col collector

cw cooling water

cwi cooling water inlet

cwo cooling water outlet

e evaporator

ex heat exchanger

ext external

g generator

HS hot stage tank

Arabic and Latin numbers correspond to states shown in Figure 1.

LITERATURE CITED

1. Versteegen P.L., and D.E. Cassel, "A survey of Existing Solar System Simulation Methods," Sun II, Proc. of ISES, Silver Jubilee Congress, Atlanta, Ga, 869 (1979)

2. Diamond S.C., H.L. Horak, B.D. Hunn, J.L. Peterson and E.F. Tucher, "DOE-1, Program Manual" Los Almos Scientific Laboratory ANL/ENG 77-04 (Oct. 1977)

3. Klein S.A., W.A. Beckman, and J.A. Duffie, "TRANSYS, A Transient Simulation Program,"ASHRAE TRANS. 82, Part 1 (1979)

4. Holmstead, "SOLSIM and Martin Marietta Interactive Thermal Analysis System-User Manual," Engineering Department Technical Manual, Martin Marietta Corp., Corp., Denver, Cal. (1976)

5. Winn C.B., G.R. Johnson and T.E. Corder, "SINSHAC: A Simulation Program for Solar Heating and Cooling of Buildings," J. of Simulation, 165 (1976)

6. ASHRAE, Handbook of Fundamentals, ch. 22 (1967)

7. Duffie, J.A., and W.A. Beckman, Solar Energy Thermal Process, John Wiley & Sons., Inc. (1974)

8. Frass A.P., and M.N. Ozisik, H Heat Exchanger Design, John Wiley & Sons, Inc. (1965)

9. Perry R.H., and C.H. Chilton, Chemical Engineers' Handbook, 5th edition, McGraw-Hill Book Company (1973)

MINERALS

15 Mineral Engineering Techniques
20 Liquid Metals Technology—Part I
43 Recent Advances in Ferrous Metallurgy
85 Fossil Hydrocarbon and Mineral Processing
173 Fundamental Aspects of Hydrometallurgical Processes
180 Spinning Wire from Molten Metals

PETROCHEMICALS

34 Petrochemicals and Petroleum Refining
49 Polymer Processing
127 Declining Domestic Reserves—Effect on Petroleum and Petrochemical Industry
135 The Petroleum/Petrochemical Industry and the Ecological Challenge
142 Optimum Use of World Petroleum

PETROLEUM PROCESSING

34 Petrochemicals and Petroleum Refining
54 Hydrocarbons from Oil Shale, Oil Sands, and Coal
98 Methanol Technology and Economics
103 C_4 Hydrocarbon Production and Distribution
127 Declining Domestic Reserves—Effect on Petroleum and Petrochemical Industry
135 The Petroleum/Petrochemical Industry and the Ecological Challenge
142 Optimum Use of World Petroleum
155 Oil Shale and Tar Sands

PHASE EQUILIBRIA

2 Pittsburgh and Houston
3 Minneapolis and Columbus
6 Collected Research Papers
81 Phase Equilibria and Related Properties
88 Phase Equilibria and Gas Mixtures Properties

PROCESS DYNAMICS

36 Process Dynamics and Control
46 Process Systems Engineering
55 Process Control and Applied Mathematics
159 Chemical Process Control

SEPARATION

91 Unusual Methods of Separation
120 Recent Advances in Separation Techniques
192 Recent Advances in Separation Techniques—II

SONICS

1 Ultrasonics—Two Symposia
109 Sonochemical Engineering

THERMODYNAMICS

7 Applied Thermodynamics
44 Thermodynamics
140 Thermodynamics—Data and Correlations

TRANSPORT PROPERTIES

16 Mass Transfer
56 Selected Topics in Transport Phenomena
77 Fundamental Research on Heat and Mass Transfer

MISCELLANEOUS

26 Chemical Engineering Education—Academic and Industrial
48 Chemical Engineering Reviews
70 Small-Scale Equipment for Chemical Engineering Laboratories
76 High Pressure Technology
112 Engineering, Chemistry, and Use of Plasma Reactors
125 Vacuum Technology at Low Temperatures
143 Standardization of Catalyst Test Methods
160 Continuous Polymerization Reactors
182 Biorheology
183 The Modern Undergraduate Laboratory Innovative Techniques
185 Electro organic Synthesis Technology
186 Plasma Chemical Processing
187 Chronic Replacement of Kidney Function
194 Hazardous Chemical—Spills and Waterborne Transportation
203 A Review of AIChE's Design Institute for Physical Property Data (DIPPR) and Worldwide Affiliated Activities
204 Tutorial Lectures in Electrochemical Engineering and Technology
206 Controlled Release Systems

MONOGRAPH SERIES

1 Reaction Kinetics by Olaf Hougen
2 Atomization and Spray Drying by W. R. Marshall, Jr.
3 The Manufacture of Nitric Acid by the Oxidation of Ammonia—The DuPont Pressure Process by Thomas H. Chilton
4 Experiences and Experiments with Process Dynamics by Joel O. Hougen
5 Present, Past, and Future Property Estimation Techniques by Robert C. Reid
6 Catalysts and Reactors by James Wei
7 The 'Calculated' Loss-of-Coolant Accident by L. J. Ybarrondo, C. W. Solbrig, H. S. Isbin
8 Understanding and Conceiving Chemical Process by C. Judson King
9 Ecosystem Technology:Theory and Practice by Aaron J. Teller
10 Fundamentals of Fire and Explosion by Daniel R. Stull
11 Lumps, Models and Kinetics in Practice by Vern W. Weekman, Jr.
12 Lectures in Atmospheric Chemistry by John H. Seinfeld
13 Advanced Process Engineering by James R. Fair